Amitkumar Mahadevan

On Row-Column-Diagonal Codes as a Class of LDPC Codes

Amitkumar Mahadevan

On Row-Column-Diagonal Codes as a Class of LDPC Codes

Properties, Decoding, and Performance Evaluation

VDM Verlag Dr. Müller

Imprint

Bibliographic information by the German National Library: The German National Library lists this publication at the German National Bibliography; detailed bibliographic information is available on the Internet at http://dnb.d-nb.de.

Cover image: www.purestockx.com

Publisher:
VDM Verlag Dr. Müller Aktiengesellschaft & Co. KG , Dudweiler Landstr. 125 a, 66123 Saarbrücken, Germany,
Phone +49 681 9100-698, Fax +49 681 9100-988,
Email: info@vdm-verlag.de

Zugl.: Baltimore, University of Maryland Baltimore County, 2005.

Produced in USA and UK by:
Lightning Source Inc., La Vergne, Tennessee, USA
Lightning Source UK Ltd., Milton Keynes, UK
BookSurge LLC, 5341 Dorchester Road, Suite 16, North Charleston, SC 29418, USA

ISBN: 978-3-639-05781-2

Curriculum Vitae

Name: Amitkumar Mahadevan.

Permanent Address: 204 Running Creek Rd., Cary, NC 27511.

Degree and date to be conferred: Doctor of Philosophy, 2005.

Date of Birth: April 3, 1977.

Place of Birth: Bombay, India.

Secondary Education:

R. N. Ruia Junior College, Bombay, India, 1994.

Indian Education Society's School, Dadar, Bombay, India, 1992.

Collegiate institutions attended:

University of Maryland, Baltimore County, Ph.D., Electrical Engineering, 2005.

University of Maryland, Baltimore County, M.S. Electrical Engineering, 2002.

Sardar Patel College of Engineering, Bombay University, Bombay, India, B.E. Electrical Engineering, 1998.

Major: Electrical Engineering.

Professional publications:

R. Holzlöhner, A. Mahadevan, C. R. Menyuk, J. M. Morris, and J. Zweck, "Evaluation of the Very Low BER of FEC Codes Using Dual Adaptive Importance Sampling", *IEEE Communications Letters*, Vol. 9, No. 2, Feb. 2005, pp. 163-165.

A. Mahadevan and J. M. Morris, "On Minimum-WER Performance of FEC Codes for the BSC/E Based on BPSK-AWGN Under Extended Bounded-Distance Decoding", under revision for *IEEE Transactions on Communications* (originally submitted May 2004).

A. Mahadevan and J. M. Morris, "SNR-Invariant Importance Sampling for Hard-Decision Decoding Performance of Linear Block Codes", submitted to *IEEE Transactions on Communications* (submitted Oct. 2004).

W. Xi, A. Mahadevan, T. Adali, J. M. Morris, "Performance Analysis of a MAP Detector for Optical Fiber Communications Systems", submitted to *IEEE Communications Letters* (submitted Jan. 2005).

A. Mahadevan and J. M. Morris, "On Quantum Codes from Weakly-Dual FEC Codes Based on Diagonals on Arrays", accepted for presentation at the *2005 Canadian Workshop on Information Theory*, Montréal, Québec, 5-8 June 2005.

A. Mahadevan, A. Mahajan, and J. M. Morris, "Some Extremely-Low WER/BER Performance Simulation Results for Linear Block Codes via DAIS", *Proceedings of the 2005 Conference on Information Sciences and Systems*, Baltimore, MD, 16-18 Mar. 2005.

A. Mahadevan, D. Mamtora, and J. M. Morris, "A Performance Surface Characterizing Sensitivity to Incorrect Channel Noise Statistics for SPA Decoding of LDPC Codes for M-QAM", *Proceedings of the 3rd International Symposium on Turbo Codes and Related Topics*, Brest, France, 1-5 Sept. 2003, pp. 559-563.

A. Mahadevan and J. M. Morris, "On FEC Code Performance Improvement Comparisons between the BSC and the BSC/E under Bounded Distance Decoding", *Proceedings of the Eighth Canadian Workshop on Information Theory*, Waterloo, Ontario, 18-21 May 2003, pp. 52-55.

A. Mahadevan and J. M. Morris, "On Minimum Probability of Error Decision Thresholds for FEC Codes on the BSC/E based on the BPSK-AWGN model", *Proceedings of the 2003 Conference on Information Sciences and Systems*, Baltimore, MD, 12-14 Mar. 2003, p. 62.

D. Mamtora, A. Mahadevan, and J. M. Morris, "On the Sensitivity of the SPA for LDPC Codes for M-QAM to Variations in Channel Noise", *Proceedings of the 2003 Conference on Information Sciences and Systems*, Baltimore, MD, 12-14 Mar. 2003, p. 63.

A. Mahadevan and J. M. Morris, "On APP Computations and Decoding Trade-offs for LDPC Codes for M-QAM", *Proceedings of the 2002 Conference on Information Sciences and Systems*, Princeton, NJ, 20-22 Mar. 2002, p. 562.

A. Mahadevan and J. M. Morris, "On LDPC Codes for ADSL", *Proceedings of the 2001 Conference on Information Sciences and Systems*, Baltimore, MD, Mar. 2001, p. 293.

A. Mahadevan and J. M. Morris, "On RCD SPC Codes as LDPC Codes Based on Arrays and Their Equivalence to Some Codes Constructed from Euclidean Geometries and Partial BIBDs", Technical Report No.: CSPL TR: 2002-1, CSPL, CSEE Dept., UMBC, Catonsville, MD, June 2002.

A. Mahadevan, S. M. Joshi and J. M. Morris, "Parameter-Error-Checking Capability added to the ADSL Simulator", Technical Report No.: CSPL TR: 2000-4, CSPL, CSEE Department, UMBC, Catonsville, MD, May 2000.

S. M. Joshi, S. Agarwal, A. Mahadevan and J. M. Morris, "ADSL Transmitter and Receiver Simulator, Ver. 0.6," Technical Report No.: CSPL TR: 2000-2, CSPL, CSEE Department, UMBC, Catonsville, MD, Apr. 2000.

Professional positions held:

Research Assistant, CSEE Department, UMBC. (Sep. 2000 - Present).

Research Intern, AT&T Shannon Labs, Florham Park, NJ. (Jun. 2000 - Aug. 2000).

Teaching Assistant, CSEE Department, UMBC. (Aug.1999 - May. 2000).

Trainee Engineer (Electrical), UHDE India Ltd., Bombay, India. (Aug. 1998 - May 1999).

Amitkumar Mahadevan, Doctor of Philosophy, 2005

Abstract

Title of Dissertation: On RCD Codes as a Class of LDPC Codes:
Properties, Decoding, and Performance
Evaluation

Amitkumar Mahadevan, Doctor of Philosophy, 2005

Advisor: Dr. Joel M. Morris
Professor of Electrical Engineering,
Department of Computer Science and Electrical Engineering

We study a class of regular low-density parity-check (LDPC) codes constructed on $\eta \times \eta$ square arrays with η prime, with the rows, columns, and diagonals of the array defining the parity-check equations. We denote such codes the row-column-diagonal (RCD) codes.

The class of RCD codes are attractive because of their high code rates coupled with the possibility of decoding via a large number of decoding techniques, as per the needs of the application. Practical decoders for RCD codes can be implemented with lower complexity than decoders for most other codes, which makes them desirable candidates for implementation in high-speed applications, e.g., optical fiber communications. RCD codes are also desirable on account of their tractability to combinatorial analysis, thanks to their relatively simple construction, which is helpful in obtaining bounds on RCD code performance in a variety of situations. Availability of these bounds, in turn, makes RCD codes excellent candidates for the testing and validation of techniques that estimate code performance.

We show that the general construction technique based on diagonals in $\eta \times \eta$ square arrays with η prime is equivalent to certain other LDPC code construction techniques based on Euclidean geometries and partial balanced incomplete-block designs. The performance of the RCD codes, which have a minimum distance of 6, is evaluated and analytically bounded (where possible) for several channel assumptions and decoding schemes: (a) the binary symmetric channel (BSC) with decoding via the bit-flipping algorithm (BFA); (b) the binary symmetric channel with erasures (BSC/E), the simplest channel model with soft-information at the channel output, with decoding based on the extended-BFA (e-BFA); and (c) the binary-input, soft-output, additive white Gaussian noise (AWGN) channel using soft-decision decoding via the sum-product algorithm (SPA).

We develop two new importance sampling (IS) simulation techniques for evaluating word-error-rate (WER)/bit-error rate (BER) performance of codes to very low values more efficiently that standard Monte Carlo techniques: (a) dual adaptive importance sampling (DAIS) for soft-decision decoding performance, and (b) SNR-invariant importance sampling (IIS) for hard-decision decoding. We provide numerous simulated WER/BER performance results for RCD codes for WERs down to 10^{-15} or lower, and validate our simulation techniques via comparison with standard Monte Carlo results (at higher WERs) and analytically computed bounds (at lower WERs). These simulation results also serve to demonstrate the behavior of the class of RCD codes and various decoders over a range of SNR values.

Finally, we demonstrate that RCD codes and codes based on diagonals in square arrays can serve as useful building blocks for constructing quantum codes that can correct for errors in the quantum domain.

Dedication

To my mother, father, and mentors

Acknowledgements

Nearly everything has changed during the course of my six-year long vigil at UMBC in my pursuit of this doctoral degree. Over this time, one of the few things that has remained unchanged has been the encouragement, support, and guidance provided by my advisor and guru Dr. Joel M. Morris. Words cannot describe the significance of his role in shaping my career. It was largely due to his persuasion that I decided to continue with my doctoral studies immediately after my masters, and boy am I glad that I took his advice! His patience, belief in my abilities (some times more than my own), and genuine desire to see me excel have had a profound impact on my research and this dissertation. He has been a constant source of motivation even on the darkest of days. I owe him my deepest gratitude for providing excellent direction to my research, unflinchingly poring through reams of drafts as well as listening to my many trial presentations and providing terrific feedback, and giving ear to my numerous and varied questions. Dr. Morris, I dedicate this dissertation in part to your unparalleled mentoring.

I am indebted to my dear uncle, Bhadran mama, who has been an immense source of support throughout my stay in the United

States. He has been a veritable friend, philosopher, and guide for me. His advice on practically all matters has had a telling influence on me and has helped me negotiate rough waters on more than one occasion. This dissertation is dedicated in part to him too. I also thank him and my aunt, Rajani mami, for having lovingly accepted me as one of their sons. It was indeed a special gesture on their part, when they drove to Baltimore from North Carolina to specifically attend my dissertation defense and provide moral support.

I thank my colleagues and lab-mates Chuck, Bill, Ron, Wen Ze, Hualiang, Dipesh, Savitha, and Anadi for having provided a stimulating environment for research, invaluable discussions, and excellent advice and criticism. Special thanks to Chuck and Bill, who in addition to Dr. Morris, have endured my numerous mock presentations (some which exceeded two hours) preceding both my proposal and final defense, and importantly for providing many useful tips and suggestions.

Many thanks are also due to the members on my defense committee, Dr. Cai, Dr. Adali, Dr. Lin, Dr. Menyuk, Dr. Pinkston, and Dr. Stephens for taking time off their busy schedules to participate in my defense, and for their useful guidance and comments. I also thank the staff of the CSEE Dept. at UMBC (Ms. Jane Gethmann and Ms. Lyn Randers in particular) for their assistance in administrativia and ever-willingness to help. I acknowledge Maryland Industrial Parterships (MIPS) / Nortel Networks (Grant No. MIPS-2218-BAY), the Department of Defense (Grant No. MDA 904-

iv

98-C-A834), the Laboratory of Telecommunications Sciences (Grant No. MDA 904-02-C0428/Z956801), and Maryland TEDCO (Grant No. P754) for partial financial support for this dissertation.

I am grateful to my roommate, Andy, for bearing with my idiosyncracies for almost my entire stay at UMBC. Special thanks my good friends Rahul and Minu, and Koustuv, Gargi, Anadi, Dipesh, Savitha, and Prajakta, to name a few others, for the many enjoyable evening and weekends, Dunkin coffee breaks, and what not. Many thanks are also due to all those cricket enthusiasts for providing me with the opportunity to continue enjoying my favorite sport.

This acknowledgement can never be complete without an expression of heartfelt gratitude for my mom and dad. I could not have come this far without their limitless love and affection. I thank them for always encouraging my educational efforts and giving me the freedom to make my decisions while at the same time not letting me go astray. They may not have been physically near me for the most part of my graduate studies, but their thoughts and well-wishes always were. I dedicate this dissertation in part to them and their supreme efforts and sacrifices.

Table of Contents

List of Tables

List of Figures

List of Acronyms

APP	:	A-Posteriori Probability
AWGN	:	Additive White Gaussian Noise
BAC	:	Binary Asymmetric Channel
BAC/AE	:	Binary Asymmetric Channel with Asymmetric Erasures
BCH	:	Bose Chaudhari Hocquenghem
BDD	:	Bounded Distance Decoding
BER	:	Bit Error Rate
BFA	:	Bit-Flipping Algorithm
BFD	:	Bit-Flipping Decoding
BIBD	:	Balanced Incomplete-Block Design
BPSK	:	Binary Phase-Shift-Keying
BSC	:	Binary Symmetric Channel
BSC/E	:	Binary Symmetric Channel with Erasures
CD	:	Constituent Decoder
CSNR	:	Channel Signal-to-Noise Ratio
CSS	:	Calderbank-Shor-Steane
DAIS	:	Dual Adaptive Importance Sampling
e-BDD	:	Extended-BDD
e-BFA	:	Extended-BFA
EG	:	Euclidean Geometry
FEC	:	Forward Error Correction
GF	:	Galois Field
HDD	:	Hard-Decision Decoding

iid	:	independent identically-distributed
IS	:	Importance Sampling
IIS	:	SNR-Invariant Importance Sampling
LDPC	:	Low-Density Parity-Check
LLR	:	Log-Likelihood Ratio
ML	:	Maximum Likelihood
MMC	:	Multicanonical Monte Carlo
NP	:	Nondeterministic Polynomial
OCR	:	Orthogonal Column Regular
OFC	:	Optical Fiber Communications
PBIBD	:	Partial Balanced Incomplete-Block Design
PCE	:	Parity Check Equation
pdf	:	probability density function
pmf	:	probability mass function
P_C	:	Probability of Correct Decoding
P_{DWC}	:	Probability of Decoding to the Wrong Codeword
P_F	:	Probability of Failure
QEC	:	Quantum Error Correction
RCD	:	Row Column Diagonal
RP	:	Random Polynomial
RS	:	Reed Solomon
SLBC	:	Systematic Linear Block Code
SNR	:	Signal-to-Noise Ratio
SPA	:	Sum-Product Algorithm
SPC	:	Single Parity Check

SPCP : Single Parity Check Product

utip : unsatisfied triple-intersection point

WE : Weight Distribution Enumerator

WEF : Weight Enumerator Function

WER : Word Error Rate

Chapter I.

INTRODUCTION

A. Overview of Dissertation

The field of error control coding has seen intense research activity since the discovery of turbo codes and decoding in 1993 [8], the rediscovery of low-density parity-check (LDPC) codes in 1997 [47], and the emergence and development of a variety of powerful, yet computationally simple, iterative techniques for decoding a large class of codes [8, 32, 41, 73, 74].

The remarkable, seemingly magical, performance achieved by these new codes and decoding techniques has hastened the entry of these new concepts into contemporary communications systems.

Furthermore, technological advancements in the communications industry have resulted in a dramatic increase in communications speeds and capacity of storage media, in addition to improvements in the overall quality of communications. This has led service providers to impose more stringent requirements on the error rate of communication systems. These tighter system requirements, in turn, have made it absolutely imperative to employ forward error correction

(FEC) techniques.

High speed communications systems such as the optical fiber communications (OFC) system usually desire low overhead, i.e., high code rate (> 0.8), error control schemes with powerful yet fast decoding algorithms [19].

It is common knowledge that the performance of FEC codes can be improved by the availability of soft-information at the output of the decision device [38,52]. This performance improvement, however, comes at the expense of an increase in the complexity and decrease in the speed of decoding, which may not be acceptable in certain scenarios.

The binary symmetric channel with erasures (BSC/E) with two thresholds and three decision levels represents the simplest example of a channel model with soft-information at its output. It has been shown [62,65] that for certain codes such as low-density parity-check (LDPC) codes, a major portion of the coding gain that can be achieved in going from hard-decision iterative decoding to completely soft-decision iterative decoding occurs in simply going to the BSC/E. The study of code performance on the BSC/E that we pursue in this dissertation is motivated by this observation.

The class of row column diagonal (RCD) single parity-check (SPC) codes (or RCD codes for short), the chief subject of this dissertation, are a family of LDPC codes with code rates approaching unity as the code length increases. These codes also lend themselves to an array of decoding schemes that provide a variety of trade-offs in terms of decoder performance on the one hand, and decoder complexity on the other. These codes, thus, are endowed with the attribute of adaptability in the face of technological improvements in decisioning and decoding techniques.

Additionally, for a given code length, the decoding complexity of message passing decoders [41], and the sum-product algorithm decoder in particular [47], is linearly proportional to the average column-weight of the LDPC code parity-check matrix. RCD codes, by virtue of having a column weight of only three – the lowest possible column weight for regular LDPC codes[1]– always result in decoders with the lowest complexity among decoders for regular LDPC codes for a given length.

RCD codes, although they have a minimum distance ($d_{\min}$) of only 6, belong to the class of weakly random-like codes [1, 55]; a desirable feature, since most successful coding schemes mimic random coding [1]. Additionally, as we will explicitly show, practical decoders for the RCD codes, such as the bit-flipping decoder, have the ability to correct a large number of error patterns of weight greater than the limit of $\lfloor \frac{d_{\min}-1}{2} \rfloor$ set by bounded-distance decoding.

Furthermore, RCD codes have an inherent structure to their parity-check matrix, as well as their codewords, which enables us to perform a combinatorial/analytical study on code properties and decoding performance.

The ability to evaluate code performance, either analytically or via simulations, to low values of bit-error rate or word-error rate is much sought after by communications engineers, especially with the advent of tighter requirements on the error rate of high-speed communication systems. Knowledge of the $d_{\min}$ of a code and the number of codewords at $d_{\min}$ ($A_{d_{\min}}$) is usually sufficient for approximating code performance under optimal maximum-likelihood decoding at high signal-to-noise ratios (low bit-error rates) via the union bound [15, 23]. For the RCD codes, this information is indeed available from the work of Martin [55],

---

[1]Gallager [27] defines regular LDPC codes to have column-weight $\gamma \geq 3$.

but for codes in general, $d_{\min}$ and $A_{d_{\min}}$ are not known.

In the absence of theoretical avenues, code performance is estimated via random simulations (standard Monte Carlo). In order to estimate very low error probabilities accurately, a large number of simulation samples need to be generated, with the number of samples being inversely proportional to the probability that needs to be estimated. Alternatively, an intelligent method that accelerates the production of error events must be available. The development of the latter idea serves as the basis for importance sampling techniques [2, 36, 42].

The ability to generate theoretical results and bounds for the performance of RCD codes makes them ideal test-candidates for studying the accuracy and capability of novel importance sampling techniques that strive to estimate code performance to very low values of error rates much below those possible via standard Monte Carlo simulations.

In this dissertation, we evaluate the performance of RCD codes using the *dual-adaptive importance-sampling* (DAIS) technique that we have developed. The DAIS evaluations for RCD codes provide estimates of error rate under the non-optimal sum-product algorithm (SPA). A comparison of these estimates against the theoretical bounds that exist for RCD codes, in turn, helps to validate DAIS. These evaluations also help us better understand the extent of the sub-optimality of the SPA decoder for RCD codes and the decoding behavior that leads to such sub-optimality.

The analysis of the decoding of error-patterns of RCD codes under BFA decoding has also led to the design of an importance sampling scheme to evaluate hard-decision decoding performance of codes in general. This importance sampling technique called *SNR-invariant importance sampling* (IIS) [54] can be interpreted as an application of the *method of stratification* [42] to the problem of evaluating hard-decision decoding performance. Under certain conditions, we show that the IIS technique is capable of evaluating code performance to arbitrarily low error-rates.

Finally, the class of codes based on diagonals in square arrays, of which the RCD codes are a special case, provide an elegant structure that helps us construct quantum error-correcting codes that are capable of correcting a class of quantum errors [16, 59].

We next present the outline for this dissertation.

B.   Outline of Dissertation

In the three chapters of this dissertation following the introduction, we describe the construction of RCD codes, discuss and analyze their decoding via the bit-flipping algorithm, and study their extension to the binary symmetric channel with erasures – the simplest example of a channel model with soft-information.

In Chapter II we present a technique to construct a family of low-density parity-check (LDPC) codes [27], based on parity-check equations defined by diagonals on a square array with sides of prime length η. The *row-column-diagonal* (RCD) *single-parity check* (SPC) codes, or RCD codes for short, are a special case of such codes. An equivalence is established between the family of LDPC codes based on diagonals in square arrays, some LDPC codes constructed from Euclidean Geometries (EG) [78], and LDPC codes constructed from lattice PBIBDs [70]. Some of the important properties of RCD codes are also discussed in this chapter.

Chapter III addresses the hard-decision iterative decoding of RCD codes on the binary symmetric channel (BSC) via the *bit-flipping algorithm* (BFA). The BFA is shown to successfully decode all error-patterns up to Hamming weight 2 for the RCD codes. A classification scheme for error-patterns of Hamming weight 3 and 4 is provided, and certain theoretical results are derived for some classes of error-patterns of weight 3. Results from an exhaustive enumeration, classification, and evaluation of 3- and 4-error patterns also are presented for a range of RCD codes. WER performance bounds for various RCD codes based on the theoretical analysis are shown and compared with the bounds based on exhaustive evaluation of 3- and 4-error patterns. All the bounds are compared also with WER performance results based on standard Monte Carlo simulations.

In Chapter IV the performance of FEC codes on the binary symmetric channel with erasures (BSC/E) is analyzed with decoding based on a simple extension to any hard-decision decoder. The improvements in code performance possible by introducing a third decision level (erasure) at the receiver are determined for a variety of code parameters under extended bounded-distance-decoding (BDD). The performance of our RCD codes is of particular interest in this study and motivates the specific choices of code parameters that are studied. The optimal location of decision thresholds is also investigated for this purpose. The performance of RCD codes under extended-BFA decoding is then addressed, and performance results for extended-BFA decoding are compared with decoding based on a different message-passing decoder for the BSC/E, as well as certain theoretical bounds.

The next two chapters are concerned with *importance sampling* techniques to evaluate the performance of codes in general, and RCD codes in particular.

The application of the dual-adaptive importance-sampling (DAIS) technique to the evaluation of low error-rates for some RCD codes under soft-decision iterative-decoding via the sum-product algorithm (SPA) in presented in Chapter V. The performance results obtained via DAIS are compared with bounds that are available for RCD codes, as well as standard Monte Carlo simulation results. Certain performance trends for RCD codes and the extent of the sub-optimality of the SPA decoder for RCD codes are discussed, along with the decoding behavior that leads to such sub-optimality.

In Chapter VI, we present an IS technique for hard-decision decoding of linear block codes for the binary symmetric channel (BSC) that exploits the concept of invariance of a hard-decision FEC decoder output to the BSC transition probability (and thus the SNR). The development of the SNR-invariant IS (IIS) method, which was inspired in part by the analysis of specific weight error-patterns of RCD codes, is presented. This sets the stage for experimentally evaluating our RCD codes for hard-decision decoding on the BSC. We analyze the IIS estimator performance in terms of its bias and variance, and obtain results from IIS simulations for some chosen codes (RCD and others). The results for RCD codes are seen to agree perfectly with the results based on the theoretical analysis and bounds presented in Chapter III.

In the penultimate chapter of this dissertation, Chapter VII, we study the application of the codes based on diagonals in square arrays to construct codes for performing error correction in the quantum domain (quantum codes). Specifically, we discuss the construction of *quantum error correction* codes from the class of classical codes based on diagonals in square arrays with η prime via the Calderbank-Shor-Steane construction technique [59]. We explicitly construct such quantum codes from diagonals based on 5×5 and 7×7 square arrays; based on observing the properties of these quantum codes, we present some heuristic design rules for the construction of such quantum codes that have a specific minimum distance and the highest possible dimensionality. We also present a technique to construct binary *weakly-dual* classical codes based on diagonals of different slopes on a rectangular $\frac{\eta}{2} \times \eta$ array (*type-R* array) with η even. Some methods to increase the minimum distance of such codes are also discussed. Quantum codes obtained from both construction techniques also are compared with some known quantum codes of similar dimensions, or with bounds (for quantum codes constructed from either classical binary codes or classical codes over GF(4)) on the highest possible minimum distance for the specified length and dimensionality.

Lastly, we end this dissertation with Chapter VIII, where we present the summary, conclusion, and directions for future research.

Chapter II.

RCD SPC CODES AS LDPC CODES BASED ON ARRAYS AND THEIR EQUIVALENCE TO SOME CODES CONSTRUCTED FROM EUCLIDEAN GEOMETRIES AND PARTIAL BIBDs

In this chapter we present a technique to construct a family of Gallager low-density parity-check (LDPC) codes [27], based on an $\eta \times \eta$ square array, where η is prime, with γ slope-q diagonal bundles representing the parity-check equations. Such codes with $\gamma = 3$ are called *row-column-diagonal* (RCD) *single-parity check* (SPC) codes. Recent research on LDPC codes has seen numerous algebraic, geometric, and combinatorial techniques proposed for their construction [37,38,70]. The primary goal of this chapter is to show that quite often these different construction techniques are indeed equivalent in terms of the codes they generate, i.e., the same code can be generated by different construction techniques. Specifically, an equivalence is established between the family of Gallager LDPC codes based on γ slope-q diagonal bundles in $\eta \times \eta$ square arrays, where η is prime, and Gallager LDPC codes constructed on Euclidean Geometries (EG) with parameters $m = 2$, $s = 1$, η and γ [38], and Gallager LDPC codes constructed from (η, γ) lattice PBIBDs [70]. We describe each of the construction techniques and demonstrate the equivalence of the three construction techniques with examples (also, see [50]). We also present some useful properties of the class of RCD codes, which will be used throughout this dissertation.

A. Gallager LDPC Codes

A regular (n, γ, ρ) LDPC code as defined by Gallager [27] is a code of block length n with each column of the parity-check matrix containing a small fixed number, γ, of ones and each row containing a small fixed number, ρ, of ones. Such codes are called LDPC codes on account of the sparseness of their parity-check matrix (**H**). In his landmark paper, Gallager only provided us with an outline for the construction of pseudo-random regular LDPC code parity-check matrices. Such codes are usually computer generated, have good minimum distance properties, and contain no short cycles of length four with high probability only if the codeword length is large and the code rate is not very high. More recently, a wide range of algebraic techniques for the construction of regular LDPC codes have been proposed [20, 37, 70, 78]. Such algebraic techniques ensure, with complete certainty, codes that have relatively good minimum distance and that do not contain short cycles of length four.

B. Gallager LDPC Codes Based on Diagonals in Square Arrays and RCD SPC Codes (Technique I)

Consider the bits in a code of length $n = \eta^2$ arranged in a square array with η bits per side. Let the rows of this array be indexed from bottom to top with $i, 0 \leq i \leq (\eta - 1)$. Similarly, we can index the columns from left to right with $j, 0 \leq j \leq (\eta - 1)$. An element in the array is then represented as $a_{i,j}$. Figure II.1 shows the labeling of elements of a square array with $\eta = 5$. A *slope-q diagonal*, $q \in \{0, 1, \ldots, \eta - 1\}$, passing through element $a_{i,j}$ in this array is a set consisting of elements $a_{\langle i+kq \rangle_\eta, \langle j+k \rangle_\eta}$, $0 \leq k \leq \eta - 1$, where $\langle b \rangle_\eta$ denotes the modulo operation $b \mod \eta$. For example, a slope-0 diagonal passing through the point $a_{i,j}$ consists of all elements in the ith row of the array, while the slope-1 diagonal passing through $a_{0,0}$ is the set $\{a_{0,0}, a_{1,1}, \cdots, a_{\eta,\eta}\}$. In addition, one also can define a slope-∞ diagonal passing through a point $a_{i,j}$ to include all elements in the jth column of the array.

Every diagonal in the array consists of exactly η elements. Further, for a fixed q, there are exactly η distinct slope-q diagonals. We define a *slope-q diagonal bundle* to be the set of all the η distinct slope-q diagonals. A slope-q diagonal bundle includes all the η^2 elements of the array with each element appearing exactly once in the bundle. Also, we have a total of $\eta + 1$ slope-q diagonal bundles, since q can take any value from the set $\{0, 1, \ldots, \eta - 1, \infty\}$. Two diagonals that belong to the same diagonal bundle do not have any element in common. Further, if η is chosen to be prime, we have an *affine plane* [67] of

$a_{4,0}$	$a_{4,1}$	$a_{4,2}$	$a_{4,3}$	$a_{4,4}$
$a_{3,0}$	$a_{3,1}$	$a_{3,2}$	$a_{3,3}$	$a_{3,4}$
$a_{2,0}$	$a_{2,1}$	$a_{2,2}$	$a_{2,3}$	$a_{2,4}$
$a_{1,0}$	$a_{1,1}$	$a_{1,2}$	$a_{1,3}$	$a_{1,4}$
$a_{0,0}$	$a_{0,1}$	$a_{0,2}$	$a_{0,3}$	$a_{0,4}$

Figure II.1: Labeling of elements of a square array with 5 elements per side

order η and it can be shown that two diagonals belonging to different diagonal bundles have exactly one element in common [10].

One defines the *incidence vector* of a particular slope-q diagonal of the array as an n-tuple with elements from GF(2). Every position in the n-tuple represents an element in the array and every element is represented exactly once. A particular position in an incidence vector of a slope-q diagonal is a 1 if and only if the element represented by that position lies on that diagonal. If an element does not belong to the diagonal, the corresponding position in the incidence vector is a 0. The construction of a code on the $\eta \times \eta$ array involves constructing parity-check equations so that they check array elements (bits) appearing in a slope-q diagonal. In particular, to construct a Gallager LDPC code with parity-check matrix column weight γ, $1 \leq \gamma \leq (\eta + 1)$, we begin by choosing any 'γ' number of slope-q diagonal bundles. The matrix formed, with the incidence vectors of the slope-q diagonals in the chosen diagonal bundles as rows, then constitutes the parity-check matrix of the code. Thus, $\mathbf{H}$ has $n = \eta^2$ columns and $\gamma\eta$ rows with every row having exactly η ones and every column having exactly γ ones. Further, if η is prime, no two different rows of the parity-check matrix have more than one 1-position in common. Hence, if η is prime, such codes do not have short cycles of length four in their factor graphs [41]. Additionally, for prime η, if we choose $\gamma = 3$ and specifically choose the slope-0 (i.e., rows), slope-∞ (i.e., columns), and slope-1 diagonal bundles to construct the parity-check matrix of a code, the resultant code corresponds to a $(\eta^2, 3, \eta)$ regular Gallager LDPC code whose minimum distance is 6 and whose factor graph has a girth of 6. We call such codes the *row-column-diagonal* (RCD) *single-parity-check* (SPC) codes with parameter η.

It is worth mentioning here that for any η, specifically choosing the rows (slope-0), columns (slope-∞) and slope-1 diagonals to construct the parity-check matrix always ensures that cycles of length-4 do not exist in the code's bipartite graph. Even values of η, however, are avoided since they lead to codes with minimum distance of 4 as opposed to a minimum distance of 6 for odd values of η (see discussion in Section II.G and [55]). Thus, the the class of RCD codes may be expanded to include all values of η (not just prime). However, the equivalence of such codes with codes constructed from other techniques can be established only for prime η.

At this point, we must state that the concept of constructing codes on arrays, with diagonals in the array defining the parity-check equations, is by no means novel. In fact, the code we have described is very closely related to *array codes* [10, 21] and codes based on *hyper-diagonal parity* [35, 72]. A significant difference between our construction technique and those presented in [10, 21] and [35, 72] is that, while our technique includes the parity-bits in the $\eta \times \eta$ array, the latter techniques operate on arrays that contain only the information bits and the parity bits are appended to the array. In the latter techniques, therefore, each of the appended parity bits participate in only one parity check equation (that parity-check equation that generates the parity bit), and the column regularity of the parity-check matrix is disrupted.

One of the first interpretations of array codes as LDPC codes was established by Fan [20]. The construction technique of [20], however, examined codewords based on a $(p-1) \times p$ array, where p was prime. Further, parity checks along the columns of the array were not considered in [20]. Given the subtle differences between the construction techniques, for codes based on diagonals in arrays, found in the literature and our construction technique, we chose to denote our codes as "codes based on diagonals in square arrays".

Finally, we also note that a technique to compute the *weight enumerator function* (WEF) [75] of RCD SPC codes has been recently established and the complete WEFs for the RCD SPC codes up to $\eta = 21$ have been evaluated [55]. Further, a combinatorial expression for the number of codewords at $d_{\min} = 6$ has also been proven, thus enabling the analytical evaluation of error-rate performance for this class of codes. The important properties of the class of RCD codes[1] are discussed in detail in Section II.G.

---

[1]Through this dissertation, RCD SPC codes may be alternatively referred to as RCD array codes or simply as RCD codes.

C. Gallager LDPC Codes Based on Lines in Euclidean Geometries (Technique II)

The construction of Gallager LDPC codes, based on the parallel structure of lines in Euclidean geometries (EG) [44], was first proposed by Xu, *et al.* [78]. The m-dimensional Euclidean geometry $EG(m, \eta^s)$ over $GF(\eta^s)$, where η is a prime, has $n = \eta^{ms}$ points, $l = \frac{(\eta^{(m-1)s})(\eta^{ms}-1)}{(\eta^s-1)}$ lines, and $\rho = \eta^s$ points on every line. Every line can be considered as either a one-dimensional subspace of the vector space of all m-tuples over $GF(\eta^s)$ or as cosets of such one-dimensional subspaces. Two lines in the EG are either disjoint, i.e., parallel, or intersect at one and only one point. The set of lines in $EG(m, \eta^s)$ can be partitioned into $K = \frac{\eta^{ms}-1}{\eta^s-1}$ sets called parallel bundles with each parallel bundle consisting of $k = \eta^{(m-1)s}$ parallel lines. Further, every parallel bundle consists of all the n points in the Euclidean Geometry with each point appearing exactly once. The incidence vector of any line in the EG is an n-tuple with elements from $GF(2)$. Every position in the n-tuple represents a point in the EG and every point is represented exactly once. A particular position in an incidence vector of a line is a 1 if and only if the point represented by that position lies on the line. If the point does not lie on the line, the corresponding point position in the incidence vector is a 0. To construct an EG Gallager LDPC code with column weight γ, $1 \leq \gamma \leq K$, we begin by choosing any 'γ' number of parallel bundles. The matrix formed with its rows as the incidence vectors of all the lines in the chosen parallel bundles then constitutes the parity-check matrix $\mathbf{H}$ of the code. Thus, $\mathbf{H}$ has n columns and γk rows with every row having exactly ρ 1s and every column having exactly γ 1s. Further, no two rows have more than one

1-position in common. Similarly, no two columns have more than one 1-position in common. Such codes, consequently, do not have short cycles of length four in their factor graphs [41].

D. Equivalence of the Codes Constructed from Techniques I and II

Consider an EG Gallager LDPC code with parameters, $m = 2$, $s = 1$, η prime, and γ parallel bundles, and a Gallager LDPC code based on a $\eta \times \eta$ array with γ diagonal bundles. In order to establish equivalence between these codes, we need to show that there exists an arrangement of the points of $EG(2, \eta)$ in a square grid of size $\eta \times \eta$ such that, for this arrangement of points, a one-to-one correspondence can be established between the lines and parallel bundles in $EG(2, \eta)$ and the slope-q diagonals and diagonal bundles in the $\eta \times \eta$ array, thus resulting in the generation of identical sets of parity-check equations.

To begin with, there are $K = \frac{\eta^{ms}-1}{\eta^s-1}$ parallel bundles in $EG(2, \eta)$ with $k = \eta^{(m-1)s} = \eta$ parallel lines in each bundle and $\rho = \eta^s = \eta$ points on each line. These numbers are exactly the same as the corresponding number of diagonal bundles, diagonals per diagonal bundle, and elements per diagonal, respectively, for the $\eta \times \eta$ array.

We now use the fact that $EG(2, \eta)$ and $GF(\eta^2)$ are isomorphic to obtain the arrangement of points in $EG(2, \eta)$ that results in equivalence of the codes. We note that every element of $GF(\eta^2)$ can be represented as a 2-tuple over $GF(\eta)$. Thus, we can arrange the points in $GF(\eta^2)$ and, hence, in $EG(2, \eta)$, in a square grid of size $\eta \times \eta$. The location of a point (x, y) in the grid is such that the

x-coordinate of the location defines the first element of the equivalent 2-tuple representation, while the y-coordinate of the location defines the second element of the 2-tuple representation.

Figure II.2 shows such an arrangement of points in $EG(2,5)$ (equivalently $GF(5^2)$) based on the primitive polynomial $x^2 + x + 2$ with coefficients from $GF(5)$. Let the root of the preceding polynomial be denoted as α, then α is a primitive of $GF(5^2)$ by definition [44]. Table II.1 lists all the elements of $GF(5^2)$ obtained from the preceding primitive polynomial and establishes the correspondence between their 2-tuple representation and the representation as powers of α (exponential representation).

Next, for the suggested arrangement of points of $EG(2, \eta)$, we use an example to demonstrate the one-to-one correspondence between the lines and parallel bundles in $EG(2, \eta)$ and the slope-q diagonals and diagonal bundles in the $\eta \times \eta$ array. Consider the Euclidean geometry $EG(2,5)$. In $EG(2,5)$, the set of points $\{0, \alpha^0, \alpha^6, \alpha^{18}, \alpha^{12}\}$ forms a line L_1. Line L_1 is also a one-dimensional subspace $((00), (01), (02), (03), (04))$ of the vector space of all 2-tuples over $GF(5)$.

A coset of L_1 called C_1 can be obtained by adding the element α^1 to each and every element of L_1 (i.e., $C_1 = \{\alpha^1 + \beta | \beta \in L_1\}$). Consequently, $C_1 = \{\alpha^1, \alpha^{17}, \alpha^{14}, \alpha^{15}, \alpha^{10}\}$. Three other distinct cosets of L_1, given by C_2, C_3, and C_4, can be obtained by adding the elements α^7, α^{19}, and α^{13}, respectively ((2 0), (3 0), and (4 0), respectively) to the elements of L_1 . Thus we get $C_2 = \{\alpha^7, \alpha^{21}, \alpha^{23}, \alpha^{16}, \alpha^{20}\}$, $C_3 = \{\alpha^{19}, \alpha^8, \alpha^4, \alpha^{11}, \alpha^9\}$, and $C_4 = \{\alpha^{13}, \alpha^{22}, \alpha^3, \alpha^2, \alpha^5\}$. Cosets C_1, C_2, C_3, and C_4 are also lines in $EG(2,5)$.

Collectively, lines L_1, C_1, C_2, C_3, and C_4 constitute a parallel bundle in $EG(2,5)$. We refer to the set $\{L_1, C_1, C_2, C_3, C_4\}$ as the parallel bundle generated

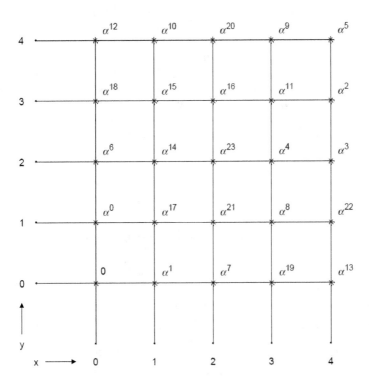

Figure II.2: Arrangement of EG$(2,5)$ points based on their 2-tuple representation over GF(5)

Table II.1

Exponential and corresponding 2-tuple representations of elements of GF(5^2) based on the primitive polynomial $x^2 + x + 2$.

Exp.	2-tuple	Exp.	2-tuple	Exp.	2-tuple	Exp.	2-tuple
$\alpha^{-\infty} = 0$	(0 0)	α^6	(0 2)	α^{13}	(4 0)	α^{20}	(2 4)
$\alpha^0 = \alpha^{24}$	(0 1)	α^7	(2 0)	α^{14}	(1 2)	α^{21}	(2 1)
α^1	(1 0)	α^8	(3 1)	α^{15}	(1 3)	α^{22}	(4 1)
α^2	(4 3)	α^9	(3 4)	α^{16}	(2 3)	α^{23}	(2 2)
α^3	(4 2)	α^{10}	(1 4)	α^{17}	(1 1)		
α^4	(3 2)	α^{11}	(3 3)	α^{18}	(0 3)		
α^5	(4 4)	α^{12}	(0 4)	α^{19}	(3 0)		

by line L_1. Also, from Figure II.2, we see that L_1 corresponds to the 0th column of the array. Similarly, C_1 , C_2, C_3, and C_4 correspond to the first, second, third, and fourth columns of the array, respectively. Clearly, one can see that the parallel bundle $\{L_1, C_1, C_2, C_3, C_4\}$ in EG(2,5) is equivalent to the slope-∞ diagonal bundle consisting of all the columns in the 5×5 array. A similar one-to-one equivalence can be established between the other parallel bundles in EG(2,5) and the diagonal bundles in the 5×5 array. In general, the equivalence between the grid arrangement of points in EG(2,η) and the $\eta \times \eta$ array holds for all prime η.

Figure II.3 shows the parity-check matrix obtained for the RCD SPC code with parameter $\eta = 5$. The labels at the top of the columns represent the locations of the η^2 square-array elements. The same parity-check matrix also is obtained if we consider a EG Gallager LDPC code with parameters, $m = 2$, $s =$

$L_R = \{0, \alpha^1, \alpha^7, \alpha^{19}, \alpha^{13}\}$ — slope-0 (rows)

$L_C = \{0, \alpha^0, \alpha^6, \alpha^{18}, \alpha^{12}\}$ — slope-∞ (columns)

$L_D = \{0, \alpha^{17}, \alpha^{23}, \alpha^{11}, \alpha^5\}$ — slope-1 (diagonals)

Figure II.3: Parity-check matrix of a $n = 25$, $\gamma = 3$, $\rho = 5$ Gallager LDPC code.

1, $\eta = 5$, and 3 parallel bundles generated by the lines $L_R = \{0, \alpha^1, \alpha^7, \alpha^{19}, \alpha^{13}\}$, $L_C = L_1 = \{0, \alpha^0, \alpha^6, \alpha^{18}, \alpha^{12}\}$, and $L_D = \{0, \alpha^{17}, \alpha^{23}, \alpha^{11}, \alpha^5\}$, respectively. The labels at the bottom of the columns represent the points of $EG(2, 5)$. Also, the labels to the right of the rows represent the slope-q diagonal bundles (in this case either the row, column, or diagonal bundles) responsible for the indicated set of parity-check equations and also the lines that generate the parallel bundles corresponding to the indicated set of parity-check equations.

Thus, for a given γ, $1 \leq \gamma \leq (\eta+1)$, the construction of Gallager LDPC codes from γ parallel bundles in $EG(2, \eta)$ is equivalent to the construction of Gallager LDPC codes from γ diagonal bundles in the $\eta \times \eta$ array. We also believe that the above result can be extended to establish equivalence between parallel bundles in $EG(m, \eta)$ and the diagonal bundles in an m-dimensional array with each side of length η, where $m > 2$.

E. Gallager LDPC Codes Based on Lattice PBIBDs (Technique III)

Gallager LDPC codes based on *partial balanced incomplete block design* (PBIBD) structures [9,11,14] constructed from integer lattices were first proposed by Vasic, *et al.* [70]. It should be noted that the generated structure is mistakenly referred to as a lattice BIBD in [70]; we will show that the structure does not satisfy the requirements of a BIBD.

Briefly, consider a set V containing v elements called points (or varieties). Let $\mathcal{B} = \{B_i : i = 1, \cdots, b\}$ be a collection of subsets of V, where each B_i contains exactly $k > 2$ elements and is called a block. If each pair of elements in V occurs together in exactly λ blocks, then $\mathcal{B}$ is called a BIBD on V [11]. The number λ is called the index of the BIBD. It can be shown [11] that every point in the BIBD appears in exactly r blocks. The parameters (v, k, λ, r, b) of the BIBD are related as follows:

$$r = \frac{\lambda \, (v - 1)}{k - 1} \tag{II.1}$$

and

$$bk = vr \; . \tag{II.2}$$

The following description is taken from [70]. Consider a rectangular integer lattice $L = \{(x, y) : 0 \leq x \leq \gamma - 1, \ 0 \leq y \leq \eta - 1\}$, where η is a prime and $\gamma \leq \eta$. Let $L \to V$ be a one-to-one mapping from the lattice L to the set of points V. Figure II.4 shows such a lattice for $\eta = 5$ and $\gamma = 3$, and for the linear mapping $l(x, y) = \eta x + y + 1$.

The blocks are generated by slope-s lines in the lattice with $0 \leq s \leq \eta - 1$, where a slope-s line passing through point (x, a) is defined to contain the points $\{(\langle x+w \rangle_\eta, \langle a+sw \rangle_\gamma) : 0 \leq w \leq \eta - 1\}$. Note that in this construction technique, slope-∞ lines are not used. Clearly, each block (from here on, we use block instead of line) has exactly γ points in it. Further, the set of all blocks with the same slope is called a *class*. There are η blocks in any given class and any class partitions the set of points V. Finally, there are exactly η classes. Thus, we have $\gamma \eta$ points, η^2 distinct blocks, γ points per block, and every point appearing in η distinct blocks.

Note that if we consider a specific point in the lattice, it appears pair-wise exactly once with every other point not in its column in the set of blocks constructed. In the set of blocks, however, that point does not appear pair-wise with

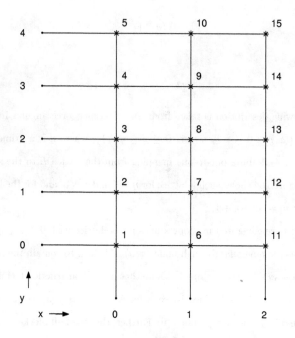

Figure II.4: Lattice of integers for $\eta = 5$ and $\gamma = 3$, and for the mapping $l(x,y) = \eta x + y + 1$.

any other point in its column. Thus, we have some pairs of points appearing once in the set of blocks, while the remaining pairs of points are not appearing even once. In this particular construction technique, consequently, the requirement that all pairs of points must appear a fixed number of times in the set of blocks is not satisfied, and we conclude that the structure that is generated is strictly not a BIBD. This fact is further confirmed by noting that equation (II.1) is not satisfied by the values of $v = \gamma\eta$, $r = \eta$, $k = \gamma$, and fixing $\lambda = 1$. Such structures are, in fact, called PBIBDs [14]; consequently, we refer to the generated structure as a (η, γ) lattice PBIBD.

The point-block incidence matrix of the (η, γ) lattice PBIBD as defined in [70] then constitutes the parity-check matrix of the code. The obtained parity-check matrix has η^2 columns and $\gamma\eta$ rows, with every row having exactly η 1s and every column having exactly γ 1s. It is shown in [70] that the Gallager LDPC codes thus constructed, have a minimum distance of 6 and do not have short cycles of length four in their factor graphs [41].

We note that the point-block incidence matrix is constructed by letting the points of the PBIBD correspond to the rows and the blocks correspond to the columns, which is the opposite of the assignment in techniques I and II where the points corresponded to the columns and the lines corresponded to the rows. This represents a fundamental difference in LDPC code construction between the first two techniques and the technique based on lattice PBIBDs.

Table II.2

Blocks of the $(5,3)$ lattice PBIBD grouped in 5 slope classes

Slope-0	Slope-1	Slope-2	Slope-3	slope-4
$b_1 = \{1,6,11\}$	$b_2 = \{1,7,13\}$	$b_3 = \{1,8,15\}$	$b_4 = \{1,9,12\}$	$b_5 = \{1,10,14\}$
$b_8 = \{3,8,13\}$	$b_9 = \{3,9,15\}$	$b_{10} = \{3,10,12\}$	$b_6 = \{3,6,14\}$	$b_7 = \{3,7,11\}$
$b_{15} = \{5,10,15\}$	$b_{11} = \{5,6,12\}$	$b_{12} = \{5,7,14\}$	$b_{13} = \{5,8,11\}$	$b_{14} = \{5,9,13\}$
$b_{17} = \{2,7,12\}$	$b_{18} = \{2,8,14\}$	$b_{19} = \{2,9,11\}$	$b_{20} = \{2,10,13\}$	$b_{16} = \{2,6,15\}$
$b_{24} = \{4,9,14\}$	$b_{25} = \{4,10,11\}$	$b_{21} = \{4,6,13\}$	$b_{22} = \{4,7,15\}$	$b_{23} = \{4,8,12\}$

F. Comparison of Gallager LDPC Codes Constructed from Techniques I and II with Codes Based on Lattice PBIBDs

When η is prime an equivalence similar to that shown in Section II.D can be established between the (η^2, γ, η) Gallager LDPC codes, constructed from γ diagonal bundles on a $\eta \times \eta$ array, and the Gallager LDPC codes based on the (η, γ) lattice PBIBDs. We specifically consider the (η, γ) lattice PBIBD, where $\eta = 5$ and $\gamma = 3$. Table II.2 lists the 25 distinct blocks grouped in five classes according to slope, for the lattice shown in Figure II.4.

The matrix shown in Figure II.3 also corresponds to the parity-check matrix of the $(5,3)$ lattice PBIBD code, where the columns correspond to the blocks as labeled in the second row below the matrix, and the rows correspond to the points of the PBIBD as labeled to the left of the matrix. The seemingly arbitrary labeling of the blocks in Table II.2 and, similarly, the seemingly arbitrary association of rows with the PBIBD points in Figure II.3, are employed with the sole intention of making the parity-check matrix of the $(5,3)$ lattice PBIBD code exactly the same as the parity-check matrices obtained from the two other construction techniques. This is perfectly acceptable since any rearrangement of the columns of a parity-check matrix resulting from a different column labeling only causes a corresponding reordering of the sequence of codeword elements, while the rearrangement of the rows of the parity-check matrix resulting from a different labeling of the rows does not have any impact on the code.

The above construction proof for establishing equivalence can be generalized to arbitrary prime η and arbitrary γ. Thus, for a given γ, $1 \leq \gamma \leq \eta$, the construction of Gallager LDPC codes from the (η, γ) lattice PBIBD is equivalent to the construction of Gallager LDPC codes from γ diagonal bundles in the $\eta \times \eta$ array.

We now briefly present some of the important properties of RCD codes that are obtained as a special case of construction technique I with $\gamma = 3$.

G. Some Properties of RCD Codes

In this section we present some properties of codes constructed from technique I, with $\gamma = 3$ and parity-check equations constructed from the slope-0 (rows), slope-∞ (columns), and slope-1 diagonal bundles. For the purpose of this section, we relax the requirement that η be prime. We shall refer to the resulting general class of codes as RCD codes.

First, we immediately observe that for arbitrary η the RCD codes are subsets of the corresponding 2-dimensional single parity-check product (SPCP) codes [12]. This follows by realizing that the 2D SPCP codes are codes constructed on $\eta \times \eta$ arrays with the parity-check equations defined by only the rows and columns of the array. For a specific η, since the codewords of the RCD codes always satisfy all the constraints that must be satisfied by codewords of the corresponding 2D SPCP code, and some additional constraints, we see that the RCD codes are a subspace of the corresponding 2D SPCP code. Mathematically, we write this as

$$C_{\text{RCD}}^{(\eta)} \subset C_{\text{SPCP}}^{(\eta)} , \tag{II.3}$$

where $C_{\text{RCD}}^{(\eta)}$ denotes the RCD code parameterized by η, and $C_{\text{SPCP}}^{(\eta)}$ the corresponding 2D SPCP.

The 2D SPCP codes have the following properties [12]:

1. Their length is $n_{\text{SPCP}}^{(\eta)} = \eta^2$ and their dimensionality is $k_{\text{SPCP}}^{(\eta)} = (\eta - 1)^2$

2. Their minimum distance is $d_{\text{SPCP}}^{(\eta)} = 4$.

3. The shortest cycles in their bipartite graphs have length 8, i.e., the 2D SPCP codes have *girth* 8.

From the properties of 2D SPCP codes and equation (II.3), we can immediately state the following properties for RCD codes:

1. Their minimum distance is $d_{\text{RCD}}^{(\eta)} \geq 4$.

2. The RCD codes have girth ≤ 8.

Note that the construction of codes based on diagonals in square arrays always enforces overall even parity on all codewords. Hence, codewords with odd Hamming weight do not exist in such codes[2].

---

[2]Summing all the row parity-check equations (PCEs), for example, gives a PCE that constrains the sum of all bits in the code to be 0 modulo 2.

Martin [55] has proved the following result for RCD codes, which is presented here without proof:

1. If η is even, $d_{\text{RCD}}^{(\eta)} = 4$.

2. If η is odd, $d_{\text{RCD}}^{(\eta)} = 6$.

Additionally, we have the following property for the dimensionality $k_{\text{RCD}}^{(\eta)}$ of RCD codes:

1. If η is even, $k_{\text{RCD}}^{(\eta)} = (\eta - 1)(\eta - 2) + 1$.

2. If η is odd, $k_{\text{RCD}}^{(\eta)} = (\eta - 1)(\eta - 2)$.

For even η, therefore, RCD codes have the parameter values

$$n = \eta^2$$
$$k = (\eta - 1)(\eta - 2) + 1 \tag{II.4}$$
$$d = 4$$

while for odd η, the RCD code parameter values are

$$n = \eta^2$$
$$k = (\eta - 1)(\eta - 2) \tag{II.5}$$
$$d = 6$$

We next prove a theorem concerning the girth of RCD codes. The following definition is needed in the proof.

Definition: A set of binary tuples is *concurring* or *orthogonal* on a set of positions (one or more) in the tuples if that set of positions has a value of unity in all the tuples and if every other position is unity in at most one tuple in the set.

If the PCEs of a code are *concurring* on each and every bit position, then the bipartite graph [41] of the code does not have 4-cycles, i.e., girth of the bipartite graph is at least 6. In simpler words, the above result may be restated as follows: For the parity-check matrix of a code, if the intersection between any two different rows (number of positions where both rows have ones) of the matrix is at most unity, then the bipartite graph corresponding to the parity-check matrix does not have 4-cycles.

Theorem II.1: *RCD codes with arbitrary η have girth = 6.*

Proof: By construction: (a) no two rows intersect each other, no two columns intersect, and no two slope-1 diagonals intersect; (b) the intersection between any row and any column is always 1; (c) the intersection between any slope-1 diagonal and any row is always 1; and, similarly, (d) the intersection between any slope-1 diagonal and any column is always 1. Hence the girth of the RCD codes is greater than 4 for arbitrary η. That the girth equals 6 can be ascertained by a simple process of locating 6-cycles on the bipartite graphs of RCD codes for any η (Note that the length of cycles is always even [11]).□

From the preceding discussion, we see that the desire to restrict η to be odd is motivated by the desire to have a minimum distance > 4; the girth is not a deciding factor in this choice.

A further restriction on η to be an odd prime may be desirable in certain cases. For one, when η is prime any two diagonals of different slopes always intersect at exactly one position and no more. If η is not prime, one must be careful in choosing the slopes to ensure intersection at only one position.

Thus, when η is prime, one may, in general, choose three different diagonal bundles of arbitrary slopes to construct a code. Such a code will be equivalent to the corresponding code constructed specifically from the slope-0, slope-1, and slope-∞ diagonal bundles, up to a permutation of codeword bit positions.

We also point out that Martin [55] has successfully shown that the number of codewords at Hamming weight 6 for an RCD code with odd η equals

$$A_6 = \eta \binom{\eta}{3}. \tag{II.6}$$

This result is extremely important from the perspective of theoretical evaluation of RCD code performance, and will be extensively used in future chapters. Finally, Martin has also shown, via its weight enumerator function, that the class of RCD codes are *weakly random-like* [1, 55], which is a desirable feature since most successful coding schemes mimic the random coding [1] weight distribution.

H. Summary and Conclusions

In this chapter we have established an equivalence between the Gallager LDPC codes constructed on a $\eta \times \eta$ square array, where η is prime, with γ slope-q diagonal bundles representing the parity-check equations, and the Gallager LDPC codes constructed based on Euclidean Geometries (EG) with parameters $m = 2$, $s = 1$, η, and γ. An equivalence between the former class of codes and Gallager LDPC codes based on (η, γ) lattice PBIBDs is also similarly established. Thus, we conclude that Gallager LDPC codes based on either of the three different construction techniques discussed eventually lead to exactly the same code.

Given these three construction techniques, one question that naturally arises is how do they compare with respect to each other. We believe that the construction technique based on diagonals in square arrays is the simplest to visualize, understand, and implement.

Since the elements of the square grid represents the codeword elements, this approach to characterize the code construction can also be helpful in the exact evaluation of the weight enumerator function (WEF) of the code, especially when $\gamma = 3$ [55]. Constructing codes using technique II involves performing much more mathematical computations as compared to technique I. Construction technique II is richer, however, since one can construct Gallager LDPC codes on an m-dimensional grid whose side is not just a prime, but also a prime power.

Construction technique III is similar to construction technique I in terms of the computations required to generate the parity-check matrix. Further, the family of codes generated using technique III (we are specifically referring to lattices with prime η and blocks constructed from parallel lines in the lattice) offers exactly the same set of code parameters as the family of codes generated using technique I. On the other hand, the approach taken in technique III to characterize the code construction may not be as helpful as technique I in the exact evaluation of the weight enumerator function (WEF) of the code when $\gamma = 3$.

The RCD codes, which are the subject of this dissertation, are a subclass of Gallager's LDPC codes, for which we presented some additional properties that will be utilized throughout this dissertation.

Chapter III.

ON BIT-FLIPPING DECODING
BEYOND THE
BOUNDED-DISTANCE DECODING
CAPABILITY FOR RCD SPC CODES

RCD codes are a family of low-density parity-check (LDPC) FEC codes that are capable of achieving high code rates. As discussed previously, these codes also lend themselves to an array of decoding schemes such as majority logic decoding, hard-decision iterative bit-flipping decoding, and soft-decision iterative sum-product decoding that provide a variety of trade-offs in terms of decoder performance on the one hand, and decoder complexity on the other. These codes, thus, are endowed with the attribute of adaptability in the face of technological improvements in decisioning and decoding techniques. These desirable features make RCD codes potential candidates for deployment in rapidly-evolving high-speed high-rate communications systems such as the optical fiber communications system.

The minimum distance $(d_{\min})$ of RCD codes can be proven to be $d_{\min} = 6$, and the number of codewords at the minimum distance $A_{d_{\min}} = \eta\binom{\eta}{3}$ [55, 56]. RCD codes, although they have a $d_{\min} = 6$, belong to the class of weakly random-like codes; a desirable feature, since most successful coding schemes mimic random coding [56]. Additionally, practical decoders for the RCD codes, such as the bit-flipping decoder, have the ability to correct a large number of error patterns of weight greater than the limit set by bounded-distance decoding. Further, RCD codes may also be employed as constituent codes in a much larger product code, with the objective of increasing length and minimum distance. RCD codes may also be used to construct codes with larger minimum-distance by employing techniques such as row-splitting [38].

LDPC codes [27] have gained in popularity as a result of the numerous efficient and powerful iterative techniques developed for their decoding. One algorithm commonly used to perform hard-decision iterative decoding of LDPC codes is the bit-flipping algorithm (BFA) [27, 38, 52]. One of the chief attributes of the BFA decoding for LDPC codes is its ability to correct a large fraction of errors beyond the code's bounded-distance error correcting capability. This chapter presents an analysis of the decoding of error patterns for the class of row-column-diagonal (RCD) codes via the BFA.

Since the RCD codes belong to the class of *orthogonal column regular* (OCR) LDPC codes, we first define OCR LDPC codes, provide some background on the BFA, and analyze the error correcting capability of the BFA for OCR LDPC codes. Based on this analysis, we show that the BFA is capable of correcting all error patterns up to Hamming weight of $\lfloor \frac{d_{\min}-1}{2} \rfloor = 2$, where $d_{\min} = 6$ is the minimum distance of RCD codes. Thus, for RCD codes, the BFA is capable of performing at least as well as a *bounded-distance decoder.*

We then analyze the ability of the BFA to decode error patterns of weight greater than the bounded-distance-decoding (BDD) capability of $\lfloor \frac{d_{min}-1}{2} \rfloor$ for the RCD SPC codes, beginning with error-patterns of weight 3. A classification scheme is developed for 3-error patterns, based on the interactions, via parity-check equations (PCEs), of the three positions in error. Important theoretical results are obtained that characterize the decoding of the defined classes and sub-classes of 3-error patterns under BFA. These results are corroborated via experimental results from the exhaustive generation, classification, and BFA decoding of 3-error patterns for RCD codes with $5 \le \eta \le 23$. Additionally, analysis of the experimental results also indicate the possibility of existence of stronger results that characterize the BFA decoding of error patterns of the various classes of weight-3.

In order to enhance the accuracy of the performance estimation over a wider range of SNRs, we also present a classification scheme for 4-error patterns that is similar to the classification of 3-error patterns. Experimental results are also obtained based on the exhaustive generation, classification, and BFA decoding of 4-error patterns for RCD codes with $5 \le \eta \le 17$.

Upper and lower bounds are computed on the word-error rate (WER) performance of certain RCD codes on a binary symmetric channel based on the theoretical results obtained for 3-error patterns. The bounds based on the theoretical results are compared with corresponding bounds from experimentally obtained results for 3-error patterns and 4-error patterns. From these performance plots it is seen that knowledge of decoding of 3-error patterns is sufficient to accurately characterize the performance of RCD codes under BFA at high signal-to-noise ratios (SNRs).

The outline of this chapter is as follows. First, we briefly discuss the BFA and the concept of orthogonal column regular (OCR) LDPC codes. Then, we show that for RCD codes, the BFA can successfully decode all error-patterns up to weight 2. Next, we present a classification scheme for 3-error patterns and prove some properties for the various classes under BFA decoding. These theoretical results help us construct a lower bound on the number of 3-error patterns that successfully decode to the all-zeros codeword for BFA decoding of RCD codes. Results from an exhaustive enumeration, classification, and evaluation of 3-error patterns for a range of RCD codes are presented next and compared with the proven theoretical results.

Similar to the 3-error patterns, we also provide a classification scheme for 4-error patterns, and present results from an exhaustive enumeration, classification, and evaluation of 4-error patterns for a range of RCD codes. WER performance bounds for various RCD codes based on the theoretical analysis are shown and compared with the bounds based on exhaustive evaluation of 3- and 4-error patterns. All the bounds are also compared with WER performance results based on standard Monte Carlo simulations. Finally, we end this chapter with the summary, conclusions, and possible areas of future research.

A. Orthogonal Column Regular LDPC Codes and the Bit-Flipping Algorithm

In this section we provide some background on OCR LDPC codes and the BFA.

1. OCR LDPC codes

Consider a binary LDPC code of length n and number of information bits k. Let the code be defined by a sparse parity-check matrix $\mathbf{H}$, having m rows and n columns, where m could be greater than $(n - k)$ as one could have redundant PCEs. We define $\mathbf{H}$ as column and row regular if every column has a fixed number γ of ones and every row has a fixed number ρ of ones, respectively.

The PCEs are said to be *concurring* or *orthogonal* on every bit position if the rows of $\mathbf{H}$ satisfy the definition of concurrency, which is as follows: A set of binary tuples is concurring on a set of positions (one or more) in the tuples if that set of positions has a value of unity in all the tuples and if every other position is unity in at most one tuple in the set. This restriction on $\mathbf{H}$ to have orthogonal PCEs on each bit position ensures that the factor graph [41] of the code does not have 4-cycles, i.e., girth of the factor graph is at least 6. Note that the requirement of the $\mathbf{H}$ matrix being row-regular can be often times relaxed without altering the analysis of the BFA. Column regularity and concurring PCEs on every bit, however, are strictly necessary. Thus, in a nutshell, OCR LDPC codes are column-regular LDPC codes with the set of PCEs concurring or orthogonal on every bit position. It can be shown (see Lemma III.1 and [44]) that the minimum distance of OCR LDPC codes with column weight γ is lower bounded by $\gamma + 1$, i.e., $d_{\min} \geq (\gamma + 1)$.

RCD codes, by virtue of their construction, are assured of having the set of row, column, and diagonal PCEs orthogonal on each and every bit-position in the codeword. Additionally, the $\mathbf{H}$ matrix of RCD codes is clearly both column and row regular, with 3 1s in every column, and η 1s in every row.

2. The bit-flipping algorithm (BFA)

The BFA was first proposed by Gallager [27] for iterative hard-decision decoding (HDD) of LDPC codes. We note that while the BFA can be applied to decode any binary linear block code, it is most effective only when the parity-check matrix is sparse, and column and row regular. Below we outline the BFA as described in [38] and [52].

Let $\mathbf{b}^{(i)} = (b_1^{(i)}, b_2^{(i)}, \ldots, b_n^{(i)})$ be the length-n binary received-word at the beginning of the ith iteration and $\mathbf{H}$ the $(m \times n)$ parity-check matrix of the code. At the first iteration, $\mathbf{b}^{(1)}$ is initialized to the received word at the output of the channel. At the ith iteration:

i) Compute the syndrome, $\mathbf{s}^{(i)} = \mathbf{b}^{(i)} \cdot \mathbf{H}^T$ (modulo-2 addition). The row vector $\mathbf{s}^{(i)} = (s_1^{(i)}, s_2^{(i)}, \ldots, s_m^{(i)})$ has length m.

ii) Compute the vector $\mathbf{d}^{(i)} = \mathbf{s}^{(i)} \cdot \mathbf{H}$ (integer addition) that denotes the number of unsatisfied PCEs for each bit in the received-word at the ith iteration. The vector $\mathbf{d}^{(i)} = (d_1^{(i)}, d_2^{(i)}, \ldots, d_n^{(i)})$ is a length-n row vector.

iii) Compute $d_{\max}^{(i)} = \max{(d_j^{(i)})}, j = 1, \ldots, n$.

iv) Compute the correction vector $\mathbf{c}^{(i)}$ as follows: For the jth position, if $d_j^{(i)} = d_{\max}^{(i)}$, then set the jth position of the correction vector to 1, i.e., set $c_j^{(i)} = 1$, else set the jth position of the correction vector to 0, i.e., set $c_j^{(i)} = 0$.

v) Compute the modified word $\mathbf{m}^{(i)}$ as $\mathbf{m}^{(i)} = \mathbf{b}^{(i)} + \mathbf{c}^{(i)}$ (modulo-2 addition).

vi) If $\mathbf{m}^{(i)}$ is a valid codeword, output $\mathbf{m}^{(i)}$ as the codeword estimate and stop, else, set $\mathbf{b}^{(i+1)} = \mathbf{m}^{(i)}$ and go to step (i).

Note that this BFA differs from Gallager's bit-flipping algorithm [27] only in step (iv). In the original bit-flipping algorithm, one flips all those bits with $d_j^{(i)} \geq T$, $j = 1, \ldots, n$, where T is a fixed predetermined threshold that does not vary from iteration to iteration. It can be shown [52] that it is possible to correct more errors by employing a threshold that equals $d_{\max}^i$ at the ith iteration than by using a fixed predetermined threshold T at every iteration. Henceforth, the term BFA identifies the algorithm described in the preceding lines and not the original bit-flipping algorithm of Gallager.

B. Channel Model and Assumptions

In this chapter, we are primarily concerned with evaluating the performance of OCR LDPC codes in general, and RCD codes in particular, under BFA decoding on a binary symmetric channel (BSC) with transition probability p. The BSC itself may be assumed to arise from a binary phase-shift-keying (BPSK) modulation scheme, a channel with additive white Gaussian noise (AWGN) of variance σ^2, and binary decisioning based on a single threshold at the receiver.

It is common practice to study the performance of linear codes by solely considering the transmission of the all-zeros codeword as a representative codeword. Such a simplification is valid provided the code is linear, and the channel and decoder satisfy certain symmetry conditions [65].

Now consider the transmission of an OCR LDPC code of length n with column weight γ on a BSC with transition probability p. It is easy to show that the BSC satisfies the symmetry conditions [65]. It was also shown in [65] that the class of *message passing decoders* satisfy the symmetry conditions. A simple comparison between the definition of the BFA and that of the the message passing decoder in [65], however, clearly indicates that the the BFA does not qualify as a message passing decoder. Even still, it turns out that decoding via the BFA can be shown to be independent of the transmitted codeword, and only dependent on the error pattern that is received. Our belief stems from the observation that the BFA requires an evaluation of the unsatisfied parity-check equations, which is clearly independent of the transmitted codeword and only dependent on the error locations. This observation can be stated in the form of a theorem, whose proof is presented next.

Theorem III.1: *The output of the BFA decoder depends entirely on the error-pattern that corrupts the transmitted codewords and does not depend on the codeword that was transmitted.*

Proof: Consider two different transmitted codewords $\mathbf{w}_1$, and $\mathbf{w}_2$, each corrupted by the error-pattern $\mathbf{e}$. Consider the computation of the syndrome $\mathbf{s}_1^{(1)}$ and $\mathbf{s}_2^{(1)}$ for the received words $\mathbf{r}_1 = \mathbf{w}_1 + \mathbf{e}$ and $\mathbf{r}_2 = \mathbf{w}_2 + \mathbf{e}$ at the first BFA iteration. Clearly $\mathbf{s}_1^{(1)} = \mathbf{r}_1 \cdot \mathbf{H}^T = \mathbf{w}_1 \cdot \mathbf{H}^T + \mathbf{e} \cdot \mathbf{H}^T = \mathbf{e} \cdot \mathbf{H}^T$, while $\mathbf{s}_2^{(1)} = \mathbf{r}_2 \cdot \mathbf{H}^T = \mathbf{w}_2 \cdot \mathbf{H}^T + \mathbf{e} \cdot \mathbf{H}^T = \mathbf{e} \cdot \mathbf{H}^T$. Thus, $\mathbf{s}_1^{(1)} = \mathbf{s}_2^{(1)}$ and we see that the syndrome computation (step (ii) of the BFA) is independent of the transmitted codeword and only depends on the error pattern $\mathbf{e}$. Further, evaluation of step (iii) of the BFA gives $\mathbf{d}_1^{(1)} = \mathbf{s}_1^{(1)} \cdot \mathbf{H} = \mathbf{d}_2^{(1)} = \mathbf{s}_2^{(1)} \cdot \mathbf{H}$. Since $\mathbf{d}_1^{(1)} = \mathbf{d}_2^{(1)}$, the bit positions that have the maximum number of unsatisfied parity-check equations associated with them are identical in both cases. This, in turn, implies that the same correction vector is applied to both $\mathbf{r}_1$ and $\mathbf{r}_2$. From the previous discussion, we see that the transmitted codeword is of little significance in determining the correction vector at the end of each BFA iteration, and this correction vector only depends on the error pattern. This preceding analysis can be iteratively applied to show that the correction vector at the end of any BFA iteration for the two received words $\mathbf{r}_1$ and $\mathbf{r}_2$ are identical. $\square$

Hence, analyzing the transmission of the all-zeros codeword captures the behavior of the BFA, and as long as the channel is symmetric, we can restrict our attention to the transmission of the all-zeros codeword, in order to study the decoding performance of the BFA.

C. Lower Bound on the Error Correcting Capability of the BFA for OCR LDPC Codes

In this section we derive a simple bound on the error correcting performance of OCR LDPC codes under BFA decoding. This bound, which is the most general bound for OCR LDPC codes, is identical to that obtained for one-step majority logic decoding [44] of one-step majority-logic decodable codes. In particular, for OCR LDPC codes with column weight γ, it has been shown that the one-step majority-logic decoder (MLD) [44] is capable of correctly decoding all error patterns with $w \le \lfloor \frac{\gamma}{2} \rfloor$. The BFA, which is essentially an iterative modification of the MLD, can be also shown capable of correctly decoding (to the all zeros codeword) all error patterns with $w \le \lfloor \frac{\gamma}{2} \rfloor$. The quantity $\lfloor \frac{\gamma}{2} \rfloor$ is often referred to as the majority-logic decoding bound. The proof of this result is presented in the following lemma.

Lemma III.1: *For an OCR LDPC code with column weight γ and the set of parity-check equations orthogonal on each and every bit position, all error patterns of Hamming weight $e \le \lfloor \frac{\gamma}{2} \rfloor$ can be corrected successfully by the BFA.*

Proof: Consider an OCR LDPC code of length n and column-weight γ, and an error pattern of length n and Hamming weight e. By the orthogonality/ concurrency condition, any non-error bit-position can participate with each of the e error positions in at most one PCE. Thus, any non-error bit-position can have at most e unsatisfied PCEs. On the other hand, an error position must have at least $\gamma - (e-1)$ unsatisfied PCEs. Such a situation occurs when the error position under consideration participates pair-wise with each of the remaining $(e-1)$ error positions in separate PCEs.

If the minimum number of unsatisfied PCEs over all error positions is greater than the maximum number of unsatisfied PCEs over all non-error positions, i.e., if $(\gamma - (e-1)) > e$, $e < \frac{\gamma+1}{2}$, or $e \leq \lfloor \frac{\gamma}{2} \rfloor$, then at the end of the first BFA iteration at least one of the error-positions will be flipped and corrected, and no new errors will be created in previous non-error positions. Thus, at the end of the first BFA iteration, the Hamming weight of the error pattern will be $e' < e$. The proof is completed by repeated application of the preceding strategy until no errors remain. $\square$

For a large class of OCR LDPC codes, which includes codes based on diagonals in square arrays and multidimensional single parity-check product codes [12], the BFA can correct all error-patterns up to a weight larger than that specified by the majority-logic decoding bound. Hence, we refer to the majority-logic decoding bound as a lower bound on the error correcting capability of OCR LDPC codes.

D. Decoding of 1- and 2-Error Patterns for RCD Codes under BFA

In this section we prove that, for RCD codes, the BFA is capable of correctly decoding all error-patterns having Hamming weight ≤ 2. As we mentioned previously in Chapter II, RCD codes constructed from the slope-0, slope-1, and slope-∞ diagonals satisfy the orthogonality condition for arbitrary η (not just prime or odd). However, a minimum distance of 6 is achieved only when η is odd. Consequently, we enforce the requirement that RCD codes have odd η. In fact, all the results proven in this chapter require η to be odd (and, thus, $d_{\min} = 6$). Henceforth, any reference to RCD codes implicitly assumes odd η unless specified otherwise.

Now consider the RCD code, which is an OCR LDPC code with $\gamma = 3$. From lemma III.1, all one-error patterns can be successfully corrected. The RCD code, however, has $d_{\min} = 6$ and, hence, a bounded-distance decoder (BDD) for the RCD code must be capable of correcting all error patterns of weight $\leq \lfloor \frac{d_{\min}-1}{2} \rfloor = 2$. An obvious question that is motivated by the preceding discussion is whether one can make any claims on the ability of the BFA to correct all 2-error patterns for the RCD codes. Demonstrating that the BFA can correct all 2-error patterns for RCD codes, in turn, would enable us to claim that the BFA can perform at least as well as the corresponding BDD. This brings us to the proof of the following theorem.

Theorem III.2: *For an RCD LDPC code of length $n = \eta^2$ (with column weight $\gamma = 3$), all error patterns of Hamming weight $e \leq 2$ can be successfully corrected by the BFA.*

Proof: From Lemma III.1, all error patterns of Hamming weight 1 can be successfully corrected. Now consider error patterns of weight-2. Let $\mathbf{b} = (b_1, b_2, \ldots, b_n)$ be the length-n binary received-word at the output of the channel. Without loss of generality, let $b_1 = b_2 = 1$ and $b_j = 0$, $j = 3, \ldots, n$. Since $\gamma = 3$, each bit-position can have at most three unsatisfied PCEs associated with it. Now consider two cases.

Case 1: Positions b_1 and b_2 do not participate together in a PCE.

In this case, each of b_1 and b_2 will have 3 unsatisfied PCEs associated with them. Any other bit-position can have at most 2 unsatisfied PCEs, thanks to the orthogonality condition. Applying the BFA to such an error pattern will correct the errors at the end of the first iteration.

Case 2: Positions b_1 and b_2 participate together in a PCE.

In this case, each of b_1 and b_2 will have $\gamma - 1 = 2$ unsatisfied PCEs associated with them because the PCE in which they participate together is satisfied. Further, there could exist other bit-positions that also have two unsatisfied PCEs. The solution to the problem lies in the answer to the question: how many other bit-positions (i.e., other than b_1 and b_2) also have two unsatisfied PCEs?

Let h_{r1}, h_{c1}, and h_{d1} denote the row, column, and diagonal PCEs in which bit-position b_1 participates. Similarly, let h_{r2}, h_{c2}, and h_{d2} denote the row, column, and diagonal PCEs in which bit-position b_2 participates. Since b_1 and b_2 participate together in one PCE, either h_{r1} and h_{r2}, h_{c1} and h_{c2}, or h_{d1} and h_{d2} must be the same. Without loss of generality, let $h_{r1} = h_{r2}$. Now, by the structure of the parity-check matrix of the RCD code, no bit-position in the code can participate simultaneously in the two PCEs h_{c1} and h_{c2} (since the column PCEs are parallel to each other and, hence, have no bit-positions in common).

Similarly, no bit-position in the code can participate simultaneously in the two PCEs h_{d1} and h_{d2}.

In order for a bit-position other than b_1 and b_2 to have two unsatisfied PCEs associated with it, that bit-position must participate with b_1 via h_{c1} and with b_2 via h_{d2} (we call this bit-position o_1), or alternatively with b_1 via h_{d1} and with b_2 via h_{c2} (we call this bit-position o_2). Clearly, exactly two distinct bit positions other than b_1 and b_2, i.e., o_1 and o_2, will have two unsatisfied PCEs associated with them. At the end of the first iteration of the BFA, errors at position b_1 and b_2 will be corrected. However, two new errors at bit-positions o_1 and o_2 will be created. Importantly, however, the number of errors does not increase at the end of the first BFA iteration.

Now the interactions between o_1 and o_2 need to be studied. If o_1 and o_2 satisfy case 1, then the errors at these bit-positions will be corrected at the end of the second iteration and no more errors will be generated, thus resulting in the successful correction of the original error pattern of weight-2 satisfying case 2. Such a situation is shown in Figure III.1. However, what if o_1 and o_2, in turn, satisfy case 2?

Let us assume that o_1 and o_2 satisfy case 2. The only way this can happen is if o_1 and o_2 lie in the same row of the RCD code, i.e., o_1 and o_2 participate together in a row PCE (say h_{r3}). (note that the assumptions of the preceding analysis and the structure of the RCD code ensures that o_1 and o_2 do not fall on the same column or diagonal.)

Now consider the set of PCEs $\Gamma = \{h_{r1}, h_{c1}, h_{c2}, h_{d1}, h_{d2}, h_{r3}\}$ (remember that $h_{r1} = h_{r2}$) and an error pattern of weight-4, with bit-positions b_1, b_2, o_1, and o_2 in error. The set Γ includes all the PCEs associated with the four bits in error.

45

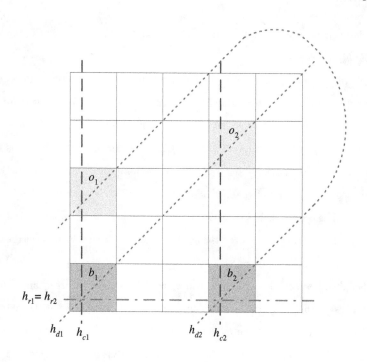

Figure III.1: Possible position of bits b_1 and b_2, and o_1 and o_2, satisfying case 2 in the proof of Theorem III.2.

Each of the PCEs in Γ has exactly two bits in error in it. Specifically h_{r1} has b_1 and b_2, h_{c1} has b_1 and o_1, h_{c2} has b_2 and o_2, h_{d1} has b_1 and o_2, h_{d2} has b_2 and o_1, and h_{r3} has o_1 and o_2. Thus, all the PCEs in Γ are satisfied. Also, the PCEs in $\mathbf{H}$, not including those in Γ, are also all satisfied since none of b_1, b_2, o_1 and o_2 participate in them. Thus, the entire parity-check matrix is satisfied by the error pattern with bit-positions b_1, b_2, o_1, and o_2 in error. In other words, the weight-4 pattern with ones located at b_1, b_2, o_1, and o_2 is not only an error pattern but also a codeword. However, the RCD code (this is where we require

η to be odd) has a minimum distance of six and, hence, cannot have a codeword of weight-4. This gives us a contradiction and, hence, the assumption that o_1 and o_2 satisfy case 2 is false.

This completes the proof of the theorem. □

The proof of the preceding theorem highlights a very important issue concerning the BFA. In general, for a given OCR LDPC code, it is unknown whether the BFA can perform at least as well as the corresponding BDD. It may be possible, however, to use the structure of the code (as done in theorem III.2) to determine whether the BFA can perform at least as well as the corresponding BDD. Finally, in addition to the RCD codes, it can also be shown by applying lemma III.1 that the BFA performs at least as well as the BDD for the class of codes that are *completely orthogonalizable* (i.e., codes where $d_{\min} = \gamma + 1$) [44].

In the next section, we specifically analyze the decoding of 3-error patterns.

E. BFA Decoding of 3-Error Patterns for RCD Codes: Classification and Analysis

In the previous section, we demonstrated (via theorem III.2) that the BFA can successfully correct all error patterns of weight-1 and weight-2 for the RCD code. The ability of the BFA to decode error patterns for RCD codes, however, is not restricted to error patterns of weight-1 and weight-2. Experiments on RCD codes have shown that error patterns of much higher weights can be decoded by the BFA (Of course, it is common knowledge that the BFA can correct patterns of weight greater than $\lfloor \frac{d_{\min}-1}{2} \rfloor$ for OCR LDPC codes in general). In this section we specifically investigate the decoding capability of the BFA for error patterns of

Hamming weight-3, henceforth called 3-error patterns, for RCD codes. The interactions via PCEs among the three different bit-positions in error will completely determine the number of unsatisfied PCEs associated with each bit-position in error. Once we determine the interactions between the different bit-positions in error and the number of unsatisfied PCEs for each of the bit-positions in error, we will need to determine how many bit-positions other than the three in error have the maximum number of unsatisfied PCEs associated with them. Since the RCD code has $d_{\min} = 6$, it would be reasonable to expect that not all 3-error patterns will be decoded successfully to the all-zeros codeword, as there will exist quite a few 3-error patterns that are also at a Hamming distance of three from some codeword with Hamming weight-6.

To facilitate the analysis of the decoding of 3-Error patterns, we first classify the 3-error patterns based on how the bit-error locations interact with each other via PCEs.

1. Classification and enumeration of 3-error patterns

Let b_1, b_2, and b_3 represent the three bit-positions in error. The primary classification of 3-error patterns is performed on the basis of the maximum number of error bit-positions that participate in a single PCE. Clearly, for a regular LDPC code in general, and for a w-error pattern, a major classification into $\min(w, \rho)$ classes can be performed (ρ is the row weight of the parity-check matrix and indicates how many bit-positions participate in any PCE), since it is possible to have anywhere between 1 and $\min(w, \rho)$ as the maximum of the error-positions participating in a single PCE.

Consequently, the major classification of the 3-error patterns for RCD codes for $\eta \geq 3$ is as follows:

Class-1:   None of the three bit-positions in error participate with each other in a PCE. In other words, there is at most only one error-position participating in any PCE.

Class-2:   There is at least one PCE in which two error positions participate, and there is no PCE that has all the three error-positions participating together.

Class-3:   All three bit-positions in error participate in the same PCE. In other words, there is at least one PCE in which all the three error-positions participate. The orthogonality condition ensures that there is exactly one such PCE in the case of RCD codes. Further, for a 3-error pattern of this class, no PCE can have exactly two participating error-positions.

In the major classification presented above, there is no further sub-classification possible in classes 1 and 3. We observe, however, that *class-2* can be further sub-classified based on how many PCEs have exactly two (of the three) participating error-positions. For regular LDPC codes, three such possibilities exist since there are $\binom{3}{2} = 3$ ways in which the three error-positions can be paired (i.e., b_1 and b_2, b_1 and b_3, and b_2 and b_3, respectively), and it is possible to have only one, any two of the three, or all three of the pairs participating via PCEs. Note that the orthogonality condition ensures that any pair can participate together in at most one PCE. Thus, the following sub-classification exists in *class-2*.

Class-2.1:   Exactly one PCE has a pair of participating error positions, and all other PCEs have at most one participating error position. Such a case arises, for example, when bit positions b_1 and b_2 participate together in a PCE, but bit positions b_1 and b_3, and b_2 and b_3 do not participate together in any PCE.

Class-2.2:   Exactly two PCEs have a pair of participating error positions, and all the other PCEs have at most one participating error position. An example is when bit positions b_1 and b_2 participate together in a PCE, and bit positions b_1 and b_3 participate together in a different PCE, but b_2 and b_3 do not participate together in any PCE.

Class-2.3:   Exactly three PCEs have a pair of participating error positions, and all the other PCEs have at most one participating error position. In other words, each of the three pairings of the error-positions participates in a distinct PCE.

Examples of 3-error patterns belonging to the different classes and subclasses for the RCD code with $\eta = 5$ are shown in Figure III.2. It is evident, based on the above classification scheme, that the classes and sub-classes partition the set of all possible 3-error patterns for LDPC codes satisfying the orthogonality condition; there will be $\binom{\eta^2}{3}$ 3-error patterns for an RCD code with array sides of length η.

It is possible to exploit the structure of RCD codes to obtain combinatorial expressions for the number of 3-error patterns belonging to each class and subclass. Consider, for example, the number of ways in which a 3-error pattern

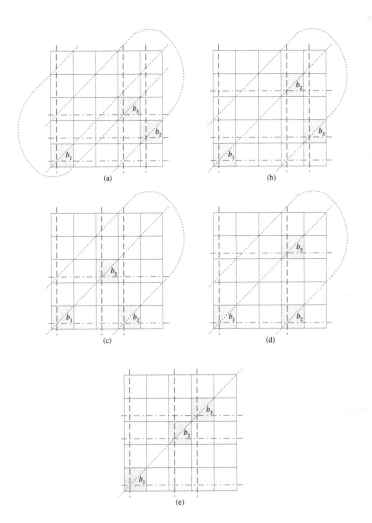

Figure III.2: 3-error patterns for the different classes and subclasses for the RCD code with $\eta = 5$: (a) *Class-1*, (b) *Class-2.1*, (c) *Class-2.2*, (d) *Class-2.3*, and (e) *Class-3*.

belonging to *class-1* can be generated for an RCD code with array sides of length η. The first error-position can be placed in any of the possible η^2 locations available in the $\eta \times \eta$ array. Once the first error position has been chosen, the second error-position must be chosen such that it does not occupy a location on any of the row, column, or diagonal PCE corresponding to the first error-position. Since each PCE of the RCD code has η participating bit-positions and the row, column, and diagonal PCEs corresponding to the first error-position have only that one bit-position in common, there are exactly $3\eta - 2$ locations in the array forbidden for insertion of the second error-position. In other words, the second error-position can be chosen from any of the remaining $(\eta^2 - 3\eta + 2)$ locations on the array.

Proceeding along similar lines, it can be shown that, for a 3-error pattern belonging to *class-1*, the third error-position can be chosen from $(\eta^2 - 6\eta + 10)$ possible positions, since the third error-position is forbidden to participate in the 6 PCEs corresponding to the first two error-positions. The quantity 10 is added since some locations on the array – those that correspond to intersections of PCEs corresponding to the first two error-positions – are counted more than once in the term 6η. Note that these expressions are obtainable via application of the *inclusion-exclusion principle* [11].

Finally, noting that the same 3-error pattern results irrespective of the order in which the three error-positions are chosen (i.e, each 3-error pattern is counted $3! = 6$ times), we obtain the following expression for the number of *class-1* 3-error patterns:

$$|C1| = \frac{\eta^2 \left(\eta^2 - 3\eta + 2\right)\left(\eta^2 - 6\eta + 10\right)}{3!} \qquad (III.1)$$

Similarly, we obtain expressions for the number of 3-error patterns belonging to the other classes and sub-classes, i.e., $|C2.1|$, $|C2.2|$, $|C2.3|$, and $|C3|$, as follows:

$$|C2.1| = \frac{\eta^2 (3\eta - 3) (\eta^2 - 5\eta + 6)}{2!} \tag{III.2}$$

$$|C2.2| = \frac{\eta^2 (3\eta - 3) (2\eta - 4)}{2!} \tag{III.3}$$

$$|C2.3| = \frac{\eta^2 (3\eta - 3) \, 2}{3!} \tag{III.4}$$

$$|C3| = \frac{\eta^2 (3\eta - 3) (\eta - 2)}{3!} \tag{III.5}$$

These expressions are consistent with the numbers obtained from an exhaustive search of all 3-error patterns for $\eta = 5, 7, 9, \cdots, 23$ (refer Table III.1).

2. Asymptotically dominant class of 3-error patterns

Based on the combinatorial expressions obtained above, it is possible to evaluate the fraction of 3-error patterns belonging to each class relative to the total number of 3-error patterns $\binom{\eta^2}{3}$. Further, we can study the behavior of such fractions as η increases and approaches infinity. This motivates the definition of the *dominant class* among the 3-error pattern classes: the dominant class of 3-error patterns is that class (or sub-class) that includes almost all the 3-error patterns in the limiting case as $\eta \to \infty$. Thus, the ratio of the cardinality of the dominant class of 3-error patterns to the total number of 3-error patterns must approach unity as $\eta \to \infty$.

Let $f_i^{(3)}$, $i \in \{1, 2.1, 2, 2, 2.3, 3\}$ denote the ratio of the 3-error-patterns in the ith class to the total number of 3-error patterns, then

$$f_1^{(3)} = \frac{|C1|}{\binom{\eta^2}{3}} = \frac{\eta^2 \left(\eta^2 - 3\eta + 2\right) \left(\eta^2 - 6\eta + 10\right)/3!}{\eta^2 \left(\eta^2 - 1\right) \left(\eta^2 - 2\right)/3!} \tag{III.6}$$

$$f_{2.1}^{(3)} = \frac{|C2.1|}{\binom{\eta^2}{3}} = \frac{\eta^2 \left(3\eta - 3\right) \left(\eta^2 - 5\eta + 6\right)/2!}{\eta^2 \left(\eta^2 - 1\right) \left(\eta^2 - 2\right)/3!} \tag{III.7}$$

$$f_{2.2}^{(3)} = \frac{|C2.2|}{\binom{\eta^2}{3}} = \frac{\eta^2 \left(3\eta - 3\right) \left(2\eta - 4\right)/2!}{\eta^2 \left(\eta^2 - 1\right) \left(\eta^2 - 2\right)/3!} \tag{III.8}$$

$$f_{2.3}^{(3)} = \frac{|C2.2|}{\binom{\eta^2}{3}} = \frac{\eta^2 \left(3\eta - 3\right) 2/3!}{\eta^2 \left(\eta^2 - 1\right) \left(\eta^2 - 2\right)/3!} \tag{III.9}$$

$$f_3^{(3)} = \frac{|C2.2|}{\binom{\eta^2}{3}} = \frac{\eta^2 \left(3\eta - 3\right) \left(\eta - 2\right)/3!}{\eta^2 \left(\eta^2 - 1\right) \left(\eta^2 - 2\right)/3!} . \tag{III.10}$$

It can be clearly seen from these expressions that $\lim_{\eta \to \infty} f_1^{(3)} = 1$, and for all other i, $\lim_{\eta \to \infty} f_i^{(3)} = 0$. This implies that *class-1* is the dominant class in the case of 3-error patterns. Further, we observe that $f_{2.1}^{(3)}$ falls off as $1/\eta$, $f_{2.2}^{(3)}$ and $f_3^{(3)}$ fall off as $1/\eta^2$, and $f_{2.3}^{(3)}$ falls off as $1/\eta^3$. This indicates that $f_{2.2}^{(3)}$, $f_{2.3}^{(3)}$, and $f_3^{(3)}$ rapidly become negligible with increasing η, but $f_{2.1}^{(3)}$ falls off more slowly and appears to be a significant contributor to the total number of 3-error patterns for low and moderate η. Figure III.3 shows the plot of $f_i^{(3)}$, $i \in \{1, 2.1, 2, 2, 2.3, 3\}$ as a function of η.

From Figure III.3, we see that at $\eta = 21$, the combined contribution of the number of 3-error patterns in *class-2.2, 2.3*, and *3* to the total number of 3-error patterns, i.e., $f_{2.2}^{(3)} + f_{2.3}^{(3)} + f_3^{(3)}$, is below 5%, and for higher values of η

this contribution decreases very rapidly. On the other hand, the contribution of *class-2.1* 3-errors patterns to the total number of 3-error patterns drops to 10% only when $\eta \approx 83$. Thus, we cannot neglect the behavior of *class-2.1* 3-error patterns for values of η in the range of current practical interest.[1]

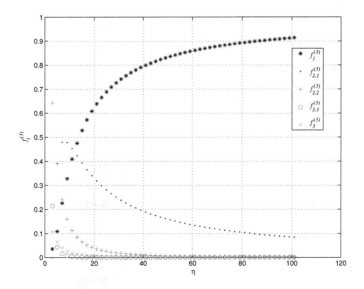

Figure III.3: Plots of $f_i^{(3)}$, $i \in \{1, 2.1, 2, 2.3, 3\}$ vs. η, with $f_1^{(3)}$ showing its dominating behavior and $f_{2.1}^{(3)}$ showing its slow fall off.

Our motivation for defining and evaluating the dominant class is that the

---

[1]Codes with lengths up to 10,000 can be considered to be of current practical interest. For longer codes, the latency in encoding and decoding may be inadmissible. Further, multiplexing of encoders and decoders may be required. Encoder and decoder complexity also increase, at least linearly, with code length.

behavior of 3-error patterns of the dominant class under the BFA is key to the overall performance of RCD codes for very large η. It would certainly be desirable if a large fraction (preferably all or almost all) of the 3-error patterns in the dominant class also decoded successfully to the all-zeros codeword. For η in the realm of practical interest, it is necessary to additionally consider the behavior of 3-error patterns of the second most dominant class, i.e., *class-2.1*. Again, having a large fraction of *class-2.1* 3-error patterns decoding successfully to the all-zeros codeword would be much desired. Ultimately we seek to characterize, as accurately as possible, the error correcting performance of RCD codes under the BFA.

In the following subsections, we prove certain results that characterize the decoding of 3-error patterns of the various classes and sub-classes.

3. Weight-6 RCD codewords and *class-1* 3-error patterns

We begin by establishing a connection between *class-1* 3-error patterns and RCD codewords at $d_{\min} = 6$. This connection eventually leads to an estimate of the number of *class-1* 3-error patterns that are not likely to decode successfully to the all-zeros codeword.

Lemma III.2: *For a weight-6 RCD codeword, any PCE can have at most two participating 1-positions.*

Proof: Every RCD codeword must have its 0- and 1-positions distributed over the array such that each and every PCE is satisfied (has even parity). Given a codeword, every PCE must have only an even number of participating 1-positions. Now, for weight-6 RCD codewords, we need to show that a PCE must have at most two participating 1-positions. We shall prove the lemma via contradiction.

Let us assume that we have a valid weight-6 RCD codeword that has a PCE with four participating 1-positions. Without loss of generality, we can assume this PCE to be a row PCE and call it r_1. Thus, there will be four locations in row r_1 of the array that will have a value of 1. Since each column of the array also defines a PCE, it follows that each of the four columns corresponding to the four ones in r_1 must have at least one more participating 1-position. Further, the columns of the array do not intersect. In other words, each of the four columns will need at least an additional distinct 1-position to satisfy its parity. It is evident that apart from the four 1-positions in r_1, at least four other distinct 1-positions are necessary to ensure that the column parities are satisfied. The Hamming weight of such a codeword must be ≥ 8, which contradicts our assumption. This completes the proof. $\square$

Lemma III.3: *Any arbitrary 1-position in a weight-6 RCD codeword must have its row, column, and diagonal parity satisfied by three additional distinct 1-positions.*

Proof: From Lemma III.2, it follows that each of the row, column, and diagonal PCE associated with a 1-position (we call it a) must have exactly one more participating 1-position. Let us call the 1-position participating in a row PCE with a as b. Similarly, c and d are the 1-positions participating in a column PCE and diagonal PCE, respectively, with a. We claim that b, c, and d are three distinct positions in the array. Again, we shall prove this lemma via contradiction.

Let us assume that any two of the three positions (say d and c) are the same. We then have bit-position a participating with bit-position $c = d$ via two separate PCEs, which is not possible since it violates the orthogonality/concurrency condition that RCD code parity-check matrices are proven to satisfy. Thus, we must conclude that $d \neq c$. Hence the proof. $\square$

Having done the groundwork, we can now proceed to prove an important result concerning weight-6 RCD codewords and 3-error patterns of *class-1*.

Theorem III.3: *Any arbitrary weight-6 RCD codeword can be partitioned into a unique pair of class-1 3-error patterns.*

Proof: Consider an arbitrary weight-6 RCD codeword and let the six 1-positions be labeled $a = (a_r, a_c)$, $b = (b_r, b_c)$, $c = (c_r, c_c)$, $d = (d_r, d_c)$, $e = (e_r, e_c)$, and $f = (f_r, f_c)$, where the subscript r and c indicate the row and column indices, respectively. From Lemmas III.2 and III.3, we know that for the considered codeword, any PCE can have at most two participating 1-positions, and each 1-position requires three additional distinct 1-positions to satisfy its row, column, and diagonal parity.

Without loss of generality, let us label the 1-position that interacts with a via its row PCE as b. We label this row PCE itself as r_1. Similarly, let us label the 1-position that participates with a via its column PCE, c_1, as c, and the 1-position that participates with a via its diagonal PCE d_1, as d. Since, each of a's PCEs are satisfied by b, c, and d, respectively, the two remaining 1-positions, e and f, cannot participate in a PCE with a. We use the notation $a \stackrel{r_1}{\sim} b$ to indicate that positions a and b participate together in PCE r_1. Similarly, we use the notation $a \nsim e$ to indicate that positions a and e do not participate together in any PCE.

Thus, based on the preceding discussion we have

$$a \overset{r_1}{\sim} b, \ a \overset{c_1}{\sim} c, \ a \overset{d_1}{\sim} d, \tag{III.11}$$

and

$$a \not\sim e, \ a \not\sim f. \tag{III.12}$$

If we can additionally prove that $e \not\sim f$, then it would follow that the three 1-positions a, e, and f constitute a *class-1* 3-error pattern. Finally, the theorem will be proved if we also show that b, c, and d constitute a *class-1* 3-error pattern, i.e., $b \not\sim c$, $b \not\sim d$, $c \not\sim d$.

We shall exhaustively consider all possible interactions between 1-positions via PCEs assuming that e and f participate together in a PCE and use the method of contradiction to establish the invalidity of these interactions. The possible interactions are depicted as a tree structure in Figure III.4.

At this point, it is appropriate to explain one of the paths of the tree depicted in Figure III.4. Since we assume that $e \sim f$, three different possibilities arise as e and f can participate together in a row, column, or diagonal PCE. These three cases represent the paths I, II, and III at the top level of the tree.

Consider path I with $e \overset{r_2}{\sim} f$. Since each 1-position must have its row, column, and diagonal PCEs satisfied, and any PCE can have at most two participating 1-positions, it follows that c and d must participate together in a row PCE, which we call r_3 (a, b, e, and f already have their row PCEs satisfied, which implies that c and d must mutually satisfy their row PCEs in order to have a valid codeword). We also note that in developing the tree, no more interactions with a are allowed, since a has all three of its PCEs satisfied.

Now, d has its row and diagonal parities satisfied by c and a, respectively, so further interaction with c and a is ruled out. Thus, three different possibilities

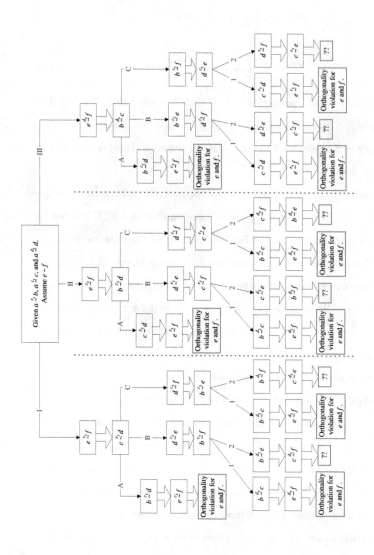

Figure III.4: Exhaustive listing of possible interactions between the 1-positions in a weight-6 RCD codeword.

(b, e, or f) arise as far as satisfying the column parity of d is concerned.[2] These three possibilities are represented as paths I.A, I.B, and I.C, respectively, in Figure III.4.

Consider path I.A. If $d \overset{c2}{\sim} b$, then it is automatically implied that e and f must also participate together in a column parity (since e and f are the only two positions remaining with unsatisfied column parity). Thus, we have $e \overset{r2}{\sim} f$ as well as $e \overset{c2}{\sim} f$, which is in direct violation of the orthogonality condition. Since RCD codes are known to meet the orthogonality condition, it follows that the interactions defined by path I.A are impossible in a RCD codeword of weight 6.

Moving on to path I.B: $d \overset{c2}{\sim} e \Rightarrow b \overset{c3}{\sim} f$, which does not violate the orthogonality condition. Thus interactions defined by this path are legal up to this point, and we must pursue the further possibilities arising from this path. Similarly, further possibilities from path I.C defined by $d \overset{c2}{\sim} f \Rightarrow b \overset{c3}{\sim} e$ must also be pursued.

In the different possibilities arising from path I.B, no further interaction with position d is allowed, since all the three PCEs of d have been satisfied. As b has its diagonal PCE still to be satisfied, two possibilities arise: $b \overset{d2}{\sim} c$, which we label as path I.B.1, and $b \overset{d2}{\sim} e$, which we label as path I.B.2 (Note that the third possibility $b \overset{d2}{\sim} f$ is not considered since it is immediately seen to violate the orthogonality condition).

Considering path I.B.1, we see that $b \overset{d2}{\sim} c \Rightarrow e \overset{d3}{\sim} f$, which is illegal on account of violating the orthogonality condition ($e \overset{r2}{\sim} f$ and $e \overset{d3}{\sim} f$). On the other hand, from path I.B.2, we have $b \overset{d2}{\sim} e \Rightarrow c \overset{d3}{\sim} f$, where the orthogonality condition is not violated.

---

[2] Of course, d may not interact with itself

At this juncture, path I.B.2 completely describes all the interactions between the 1-positions of a weight-6 RCD codeword. In order to resolve the validity of the codeword defined by path I.B.2 we must determine whether or not it is physically possible to construct a weight-6 RCD codeword with interactions defined by path I.B.2. In other words, we need to ask and answer the question: Is there some other code property that explicitly prevents us from physically realizing a codeword defined by the interactions in path I.B.2. Given our lack of knowledge, we terminate this path by a box with the ?? sign that indicates that further analysis needs to be performed on the validity of this path.

When the tree structure representing the possible interactions is completed (as shown in Figure III.4), we see that all but six paths describe impossible interactions among the 1-positions in a weight-6 RCD codeword. Close observation of these six paths that survive (the paths terminated by a box with the ?? sign) reveals that they are equivalent to each other up to a relabeling of 1-positions and parity-check equations. For example, it can be seen that paths I.B.2 and I.C.2 describe the same interactions, with the only difference being that the labels for positions e and f are swapped. Thus, it is sufficient to only investigate the validity of the interactions defined by path I.B.2. The result of this investigation can then be extended to the other paths almost trivially.

We emphasize that until this point, we have only used the orthogonality condition to invalidate some of the possible interactions indicated in the tree of Figure III.4. The orthogonality condition by itself is insufficient to eliminate the six surviving paths. Hence, we now need to invoke the properties of the geometric structure of the RCD code to perform further analysis. In particular, the benefit of restricting η to an odd value will become apparent in the following analysis, since it is precisely this constraint that we exploit in order to invalidate the existence of each of the six surviving paths and, thus, complete our proof.

The interactions defined by the PCEs of path I.B.2 describe a set of linear constraints on the row and column indices of the six 1-positions.[3] In particular, based on only the row and column PCEs we have

$$
\begin{aligned}
a \overset{r_1}{\sim} b &\;\Rightarrow\; a_r = b_r \\
a \overset{c_1}{\sim} c &\;\Rightarrow\; a_c = c_c \\
e \overset{r_2}{\sim} f &\;\Rightarrow\; e_r = f_r \\
c \overset{r_3}{\sim} d &\;\Rightarrow\; c_r = d_r \\
d \overset{c_2}{\sim} e &\;\Rightarrow\; d_c = e_c \\
b \overset{c_2}{\sim} f &\;\Rightarrow\; b_c = f_c
\end{aligned}
\tag{III.13}
$$

From equation (III.13) it is clear that, based on the row and column constraints, only six of the twelve index variables corresponding to the six 1-positions are independent. We arbitrarily choose a_r, a_c, d_r, d_c, f_r, and f_c as the independent index variables.

These six independent variables are additionally constrained by the diagonal PCEs and, further, we also have the requirement that

$$
\begin{aligned}
a_r \neq d_r \neq f_r \neq a_r \\
a_c \neq d_c \neq f_c \neq a_c
\end{aligned}
\tag{III.14}
$$

(if not, then we will have PCEs with more than two participating 1-positions).

---

[3]Remember that the row and column indices are constrained to take on integer values from the set $\{0, 1, \ldots, \eta - 1\}$.

The constraints corresponding to the diagonal PCEs can be written as

$$a \overset{d_1}{\sim} d \quad \Rightarrow \quad a_r - a_c = (d_r - d_c) \bmod \eta$$

$$\text{or} \quad a_r - a_c - d_r + d_c = 0 \bmod \eta \qquad \text{(III.15)}$$

$$b \overset{d_2}{\sim} e \quad \Rightarrow \quad b_r - b_c = (e_r - e_c) \bmod \eta$$

$$\text{or} \quad a_r - f_c - f_r + d_c = 0 \bmod \eta \qquad \text{(III.16)}$$

$$c \overset{d_3}{\sim} f \quad \Rightarrow \quad c_r - c_c = (f_r - f_c) \bmod \eta$$

$$\text{or} \quad d_r - a_c - f_r + f_c = 0 \bmod \eta, \qquad \text{(III.17)}$$

where equations (III.16) and (III.17) are obtained by substituting the independent index variables for the dependent index variable based on equation (III.13).

Finally, since the diagonal PCE in which a pair of 1-positions (a and d, or b and e, or c and f) participate must be different from the diagonal PCE of the other pairs, we obtain an additional set of constraints, which are given as

$$a_r - a_c \neq (f_r - f_c) \bmod \eta \qquad \text{(III.18)}$$

$$a_r - a_c \neq (f_r - d_c) \bmod \eta \qquad \text{(III.19)}$$

$$f_r - f_c \neq (f_r - d_c) \bmod \eta \qquad \text{(III.20)}$$

Equations (III.14) to (III.20) represent a set of constraints that the six independent index variables must satisfy. Our task then is to determine whether or not this set of equations has solutions when the index variables are further constrained to take values from the set $\{0, 1, \ldots, \eta - 1\}$.

Subtracting equation (III.16) from equation (III.15) gives us

$$-a_c - d_r + f_r + f_c = 0 \bmod \eta. \qquad \text{(III.21)}$$

Further, subtracting equation (III.21) from equation (III.17) gives us

$$2d_r - 2f_r = 0 \bmod \eta. \qquad \text{(III.22)}$$

Now remembering that $d_r, f_r \in \{0, 1, \dots, \eta - 1\}$, we get

$$-2(\eta - 1) \le 2(d_r - f_r) \le 2(\eta - 1). \qquad \text{(III.23)}$$

We now use the fact that η must be odd (based on the definition of RCD codes) to show that there exist no index variables that can satisfy the constraints defined by equations (III.14) to (III.20). Since η is odd and $2(d_r - f_r)$ is necessarily even, if equation (III.22) has to be satisfied, we must have

$$2(d_r - f_r) \in \{\dots, -4\eta, -2\eta, 0, 2\eta, 4\eta, \dots\}. \qquad \text{(III.24)}$$

Combining the requirements imposed by (III.23) and (III.24), we see that the only common solution is given by $2(d_r - f_r) = 0$ or $d_r = f_r$. This directly contradicts, however, the requirement imposed by equation (III.14). Thus, we have a contradiction, which implies that the set of constraints described by equations (III.14) to (III.20) has no solution in the set $\{0, 1, \dots, \eta - 1\}$ when η is odd.

It follows that the interactions defined by path I.B.2 (and the other five survivors from Figure III.4) are not physically realizable in weight-6 RCD codewords. Thus, we must conclude that $e \nsim f$ and that a, e, and f constitute a *class-1* 3-error pattern.

To show that b, c, and d also constitute a *class-1* 3-error pattern, it is sufficient to observe that requiring any two of b, c, and d to participate together in a PCE automatically necessitates e and f to also participate together in a PCE. For example, assuming $b \overset{d}{\sim} c$ implies that $e \overset{d}{\sim} f$. Note that since $a \overset{r}{\sim} b$, $a \overset{c}{\sim} c$, and $a \overset{d}{\sim} d$, b can participate with c only in a diagonal PCE. Similarly, b can participate with d only in a column PCE, which would imply $e \overset{c}{\sim} f$. Thus, b, c, and d also constitute a *class-1* 3-error pattern.

Finally, the uniqueness of the partition of a weight-6 RCD codeword, in terms of a pair of *class-1* 3-error patterns, follows from the construction. □

We see from theorem III.3 that we have $\leq 2A_{d_{\min}}$ *class-1* 3-error patterns that are equidistant from the all-zeros codeword and a codeword at $d_{\min} = 6$, with equality being achieved if any two different codewords at $d_{\min} = 6$ result in non-identical *class-1* 3-error pattern partition elements. Thus, at the most, $2A_{d_{\min}}$ *class-1* 3-error patterns are likely to not decode successfully to the all-zeros codeword.

To further understand why those *class-1* 3-error patterns that partition an RCD codeword at $d_{\min} = 6$ are likely to not successfully decode to the all-zeros codeword, it is useful to introduce the following definition.

Definition: An *unsatisfied triple intersection point* (*utip*) for an RCD code error-pattern is defined as a location on the RCD code array that has all three of its associated PCEs unsatisfied.

We immediately see that for a *class-1* 3-error pattern, each of the three positions in error is indeed a *utip*. Further, a *class-1* 3-error pattern that partitions an RCD codeword at $d_{\min} = 6$ (say the pattern with ones at positions a, e, and f in the proof of Theorem III.3) has at least three additional *utips* defined by the other partition element (i.e., the pattern with ones at positions b, c, and d in the proof of Theorem III.3).

Assuming, for a moment that there are no more *utips* other that those already indicated, we see that at the first BFA iteration during the decoding of the *class-1* 3-error pattern defined by positions a, e, and f, the six positions a, e, f, and b, c, d, will each have three unsatisfied PCEs associated with them. Thus, each of these six positions will be flipped at the end of the first BFA iteration. While all three original errors (a, e, and f) are corrected, three new errors (at positions b, c, and d) are introduced. Of course, from Theorem III.3, the three new errors constitute a *class-1* 3-error pattern that partitions an RCD codeword of weight-$d_{\min} = 6$. Now each of the three positions b, c, and d is a *utip*. Simultaneously, the positions a, e, and f also constitute three additional *utips*.

At the second BFA iteration, the same six positions b, c, d and a, e, f will each again have three unsatisfied PCEs associated with them. Thus all six positions will be flipped at the end of the second BFA iteration. The new error pattern at the end of the second BFA iteration, thus, is identical to the original error pattern. The output of the BFA, therefore, toggles between two *class-1* 3-error patterns from iteration to iteration and fails to converge to a valid codeword. Consequently, we have a decoder failure.

The preceding discussion also leads us to conclude that if a *class-1* 3-error pattern has two or fewer additional *utips*, then it is guaranteed to decode successfully to the all-zeros codeword. This follows by observing that all *utips*, including the three original error-positions themselves, are flipped at the end of the first BFA iteration. Thus all the original errors are corrected. At the same time at most two new errors corresponding to the additional *utip* are introduced. The error-pattern at the end of the first BFA iteration has Hamming weight 2 or less and, hence, based on Theorem III.2, is successfully decoded to the all-zeros codeword.

4. *Class-2.2* 3-error patterns and their decoding via BFA

In this subsection, we prove an important result concerning 3-error patterns belonging to *class-2.2*. Exhaustive enumeration and decoding via BFA of 3-error patterns for RCD codes[4] over a range of values of η has led to the conjecture that exactly $3\eta^2 (\eta - 1)$ *class-2.2* 3-error patterns get decoded to a codeword other than the all-zeros, while decoding of the rest of the *class-2.2* 3-error patterns results in decoder failure. Further, we have observed that whenever a non-all-zeros codeword results from BFA decoding of a *class-2.2* 3-error pattern, it always has a Hamming weight of 2η, and the decoding is completed in exactly two BFA iterations. In the following theorem, we provide a proof for this phenomenon.

We begin with a discussion on the two different types of *class-2.2* 3-error patterns. Figure III.5 shows two different *class-2.2* 3-error patterns for the RCD code with $\eta = 5$. In both cases the 1-positions are indicated by labels e_1, e_2, and e_3. The pattern in Figure III.5(a) has exactly one *utip* indicated by the position o_1. We call this a *category-1 class-2.2* 3-error pattern. On the other hand, the pattern in Figure III.5(b) has no *utip*: we call this a *category-0 class-2.2* 3-error

---

[4]These results are presented in Section III.F

pattern.

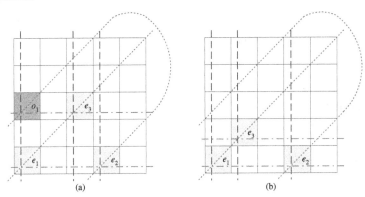

(a)   (b)

Figure III.5: Two different categories of *class-2.2* 3-error patterns: (a) *category-1* and (b) *category-0*.

The exact number of *category-1 class-2.2* 3-error patterns can be ascertained by construction, which leads us to the following lemma.

Lemma III.4: *There are exactly $3\eta^2 (\eta - 1)$ class-2.2 3-error patterns that have a single utip and, thus, belong to category-1.*

Proof: We shall prove this lemma via construction. Consider an RCD code array with side of length η. Let us label the 1-position that interacts with two other 1-positions separately as e_1. The two other 1-positions are labeled e_2 and e_3. Based on the previously defined notation, we must have $e_1 \sim e_2$, $e_1 \sim e_3$, and $e_2 \nsim e_3$ (Figure III.5 gives an example for the specific case where $\eta = 5$).

Without loss of generality, we also assume that $e_1 \overset{r}{\sim} e_2$ and $e_1 \overset{d}{\sim} e_3$. It is worthwhile noting that an *utip*, if it exists, must fall on the PCE of e_1 that does not include the interaction with either e_2 or e_3. Thus, in the case under consideration, the *utip* must fall on the column PCE of e_1 (the row and diagonal

PCEs of e_1 are satisfied). Now, if we consider e_2 we can similarly infer that the *utip*, if it exists, must fall on the diagonal of e_2 (the row of e_2 is satisfied, and the column of e_2 can never intersect with the column of e_1). Analogously, the *utip*, if it exists, must fall on the row of e_3.

The preceding observations provide us with sufficient information to develop a recipe to construct a *category-1 class-2.2* 3-error pattern. The steps are as follows:

1. Choose a position for e_1. There are η^2 different ways in which this can be done.

2. Choose two out of the three PCEs of e_1. The 1-positions e_2 and e_3 will be placed separately on each of the chosen PCEs. There are $\binom{3}{2} = 3$ different possibilities here.

3. Choose a specific position for e_2 on one of the two chosen PCEs of e_1. There are $(\eta - 1)$ different possibilities here.

4. The position of e_3 is automatically determined on the second chosen PCE of e_1 that leads to an *utip*.

It remains to verify that there is always exactly one position that e_3 can take once e_2 has been determined, and given this, that e_2 and e_3 themselves do not interact via a common PCE (thus resulting in a *class-2.3* 3-error pattern instead of a *class-2.2* 3-error pattern). To ascertain the first part, we shall make use of the specific example we constructed. The inferences drawn from this example are readily generalized.

Consider the construction of the *class-2.2* 3-error pattern in Figure III.5(a). We first choose a position for e_1 and then choose its row and diagonal PCEs to place e_2 and e_3, respectively. Further, once we specify the location of e_2, the potential position of an *utip* is automatically revealed at o_1, the intersection of the column of e_1 and the diagonal of e_2. Exactly one potential *utip* always exists, since any two non-parallel PCEs in an RCD code always intersect at one position. The row of o_1 will intersect the diagonal of e_1 at exactly one point that then defines the location of e_3.

Now it only remains to show that $e_2 \nsim e_3$. Let us assume that $e_2 \sim e_3$. We observe that for this to happen, the PCE in which e_2 and e_3 participate together must be a column PCE (it cannot be a row PCE since $e_1 \overset{r}{\sim} e_2$ and it cannot be a diagonal PCE since $e_1 \overset{d}{\sim} e_3$). If we now consider the four 1-positions e_1, e_2, e_3, and o_1, we observe that these four positions mutually satisfy their PCEs and must constitute a valid codeword of weight 4. Happily though, weight-4 codewords do not exist in RCD codes resulting in a contradiction. We must conclude, therefore, that $e_2 \nsim e_3$.

The previous arguments are easily seen to hold for all values of odd η. Therefore, based on the preceding discussion, there exist exactly $3\eta^2 (\eta - 1)$ *class-2.2* 3-error patterns that each have a single *utip* and, thus, belong to *category-1*. $\square$

Next, we state and prove the main theorem of this section.

Theorem III.4: *Exactly $3\eta^2 (\eta - 1)$ class-2.2 3-error patterns each have a single utip, and each of these decodes to a codeword of Hamming weight 2η, thus generating a decoder error in exactly 2 BFA iterations.*

Proof: The first part of the theorem directly follows from Lemma III.4. We need to investigate the BFA decoding of a *category-1 class-2.2* 3-error pattern to prove the second part.

We shall use the general structure of a *category-1 class-2.2* 3-error pattern in the steps of this proof without making reference to any specific value of η. We do provide an example to assist in the comprehension of the proof, but, as will be evident, the following discussion is general for the most part, and any specific reference to the illustrated example trivially generalizes to all *category-1 class-2.2* 3-error patterns.

Consider the BFA decoding of a *category-1 class-2.2* 3-error pattern (say the one in Figure III.5(a)). Such an error-pattern has exactly one *utip* denoted by position o_1, while e_1, e_2, and e_3 represent the three 1-positions that are assumed to satisfy (without loss of generality) $e_2 \sim e_3$, $e_1 \overset{r}{\sim} e_2$, and $e_1 \overset{d}{\sim} e_3$. Figure III.6 shows the different steps of BFA decoding of the error-pattern in Figure III.5(a). In Figure III.6(a) the unsatisfied PCEs are indicated via lines on the array, and the number in each element of the array indicates the number of unsatisfied PCEs associated with that bit-position. The array elements filled with a light shade of grey represent the 1-positions in the error-pattern. All other bit-positions have value zero. The only bit-position that has three unsatisfied PCEs is the *utip* o_1 (indicated by a dark shade of grey).

At the end of the first BFA iteration, the bit-positions having the maximum number of unsatisfied PCEs are flipped. Clearly, for all *category-1 class-2.2* 3-error patterns (and for all η odd), the only bit-position that will be flipped (more specifically converted to 0 from 1) will be the *utip*. The resulting error-pattern now has Hamming weight 4 and is shown in Figure III.6(b) (for the specific example being considered) along with the PCEs that now become unsatisfied.

We see that only two PCEs will remain unsatisfied. Further, these two PCEs with participating 1-positions e_2 and e_3, respectively, are parallel to each other. They are also parallel to that PCE of e_1 that does not contain either e_2 or

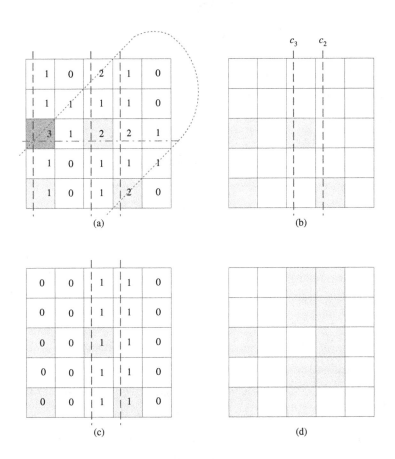

Figure III.6: BFA decoding of a *category-1 class-2.2* 3-error pattern for the RCD code with $\eta = 5$.

e_3 (in the example, the column PCE of e_1 does not contain either e_2 or e_3). Consequently, the two unsatisfied PCEs at the end of the first BFA iteration are the column PCEs passing through e_2 and e_3.

It can be easily verified based on the geometric structure of the weight-4 error-pattern with 1-positions given by e_1, e_2, e_3, and o_1 that all PCEs other than the two described previously are satisfied. Thus, during the second BFA iteration, the bit-positions on each of the two unsatisfied PCEs (which include e_2 and e_3) have a single unsatisfied PCE associated with them. Let us label these two unsatisfied PCEs as c_2 and c_3, respectively. All other bit-positions have zero associated unsatisfied PCEs. This situation is illustrated in Figure III.6(c) for the example on hand, where again, the 1-positions in the error-pattern are indicated by the light shade of grey.

At the end of the second BFA iteration, all bit-positions on c_2 and c_3 are flipped. More specifically, e_2 and e_3 will be converted from 1 to 0, while all other bit-positions on c_2 and c_3 are converted from 0 to 1. The resulting error-pattern will have a Hamming weight of 2η, since there are $\eta - 1$ 1-positions on each of c_2 and c_3, and two more ones at locations e_1 and o_1. For our example, this error-pattern is shown in Figure III.6(d).

All that remains to be shown is that the error-pattern after the second BFA iteration is in fact a valid codeword. First, we observe that c_2 and c_3 each have even parity since η is odd. The PCE c_1 that is parallel to c_2 and c_3, and passes through e_1, also contains o_1 by construction, and hence, is satisfied. Every other PCE parallel to c_2 and c_3 has zero participating 1-positions and, hence, is satisfied. Thus all PCEs in the diagonal bundle that contains c_2 and c_3 are satisfied.

Next, we turn our attention to the other two parallel bundles (without loss of generality, we assume them to be the rows and diagonals, which is in accordance with our example). Each row PCE will intersect both c_2 and c_3 at single positions. By construction the row of e_1 contains e_2. We label this row PCE r_1. We also see that the row of o_1 must contain e_3. We label this row PCE r_2. Barring these two row PCEs, all other row PCEs have exactly two 1-positions in them, one contributed by c_2 and the other contributed by c_3, and, thus, are satisfied.

If we consider the row PCE r_1, we observe that again there are exactly two 1-positions on it, given by e_1 and the intersection of r_1 and c_3. Note that the position e_2 no longer has a value of 1, since it was flipped at the end of the second BFA iteration. Similarly, row PCE r_2 contains exactly two 1-positions given by o_1 and the intersection of r_2 and c_2. Thus both r_1 and r_2 are satisfied, which implies that all PCEs in the row parallel bundle are satisfied.

The parallel bundle of diagonal PCEs can be analyzed in a similar fashion to that of the row parallel bundle to verify that all PCEs in the diagonal parallel bundle are also satisfied. Thus, the error pattern at the end of the second BFA iteration is indeed a valid codeword and constitutes a decoder error. Combining all intermediate results completes the proof. □

5. *Class-2.1* 3-error patterns and their decoding via BFA

Similar to the *class-2.2* 3-error patterns that were analyzed in the preceding section, much can be said about BFA decoding of *class-2.1* 3-error patterns. Our exhaustive enumeration and decoding of this class for a range of values of η has shown that the constituent error-patterns either decode successfully to the all-zeros codeword or lead to decoder failure. However, no error-pattern of this class gets decoded to a wrong codeword (see Table III.1).

Based on the statistics of the exhaustive experiments, we can also conjecture that $|C2.1| - 3\eta^2 (\eta - 1) (\eta - 3)$ *class-2.1* 3-error patterns decode successfully to the all-zeros codeword, where $|C2.1| = \frac{3}{2}\eta^2 (\eta - 1) (\eta^2 - 5\eta + 6)$ is the total number of *class-2.1* 3-error patterns.

We commence the analysis of this class by presenting a few lemmas.

Lemma III.5: *A class-2.1 3-error pattern that has one of the error-positions as its only utip will always decode successfully to the all-zeros codeword.*

Proof: First, by definition, a *class-2.1* 3-error pattern will have one error-position that does not participate together via PCEs with any of the other two error-positions. Let us call this error-position e_1. The other two error-positions, denoted by e_2 and e_3, must then participate together in a PCE. Clearly, e_1 will be a *utip*. If we assume that the error-pattern has no *utip* other than e_1, then we observe that at the end of the first BFA iteration, the value of e_1 will be flipped and become a 0. Additionally, no other bit-position will be flipped. Thus, at the end of the first BFA iteration, we will be left with an error-pattern of Hamming weight 2. Since all error-patterns of Hamming weight 2 are successfully decoded to the all-zeros codeword (refer Theorem III.2), the proof is complete. □

Lemma III.6: *A class-2.1 3-error pattern with error-positions e_1, e_2, and e_3, where the positions are labeled to satisfy $e_1 \nsim e_2$, $e_1 \nsim e_3$, can only have either zero or exactly one utip other than e_1. Further, if one additional utip exists, it must fall on that PCE of e_1, which is parallel to the PCE that contains e_2 and e_3.*

Proof: We shall prove this lemma via construction. Consider the following steps in the construction of an arbitrary *class-2.1* 3-error pattern.

1. Choose a position for e_1. There are η^2 different ways in which this can be done.

2. The two other 1-positions, e_2 and e_3, cannot lie on any of the three PCEs associated with e_1. At the same time, e_2 and e_3 must participate together in a PCE. Thus we can choose a PCE on which to locate e_2 and e_3 from among $3\eta - 3$ available PCEs in $3(\eta - 1)$ ways. Based on the construction up to this point, it is obvious that the position e_1 must be a *utip*.

3. Once the PCE that must contain e_2 and e_3 has been chosen, it remains to determine the different possible ways in which e_2 and e_3 may be positioned and the impact of the positioning on the existence of additional *utips*.

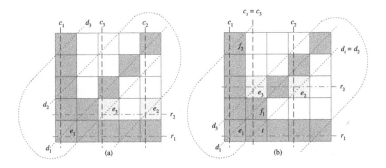

Figure III.7: *Class-2.1* 3-error patterns for the RCD code with $\eta = 5$.

Any PCE apart from the three associated with e_1 will have $(\eta - 2)$ valid positions in which e_2 and e_3 can be located. The two forbidden positions correspond to the intersection of the chosen PCE with each of the two PCEs associated with e_1 that are not parallel to the chosen PCE. Figure III.7(a) provides an example of an arbitrary 3-error pattern belonging to *class-2.1* for the RCD code with $\eta = 5$.

In Figure III.7(a), the three PCEs r_1, c_1, and d_1, associated with e_1, are indicated by lines on the array[5], and the forbidden positions for placing e_2 and e_3 are indicated by elements having a dark shade of gray.

Without loss of generality, let us assume that we choose a row PCE, which we label as r_2, to place the error-positions e_2 and e_3, and let c_2 and d_2 denote the column and diagonal PCEs of e_2, respectively, and c_3 and d_3 denote the column and diagonal PCEs of e_3, respectively.

For an *utip* other than e_1 to exist, we require three unsatisfied non-parallel PCEs to intersect at a common point. For the *class-2.1* 3-error pattern under consideration, only one row PCE, r_1, is unsatisfied (since r_2 is satisfied by construction because it has both e_2 and e_3 participating). It follows that any additional *utip*, if it exists, must lie on r_1. Thus there are two possible candidates for the additional *utip*. The first is the hypothetical point defined by the intersection of r_1, d_2, and c_3. Let us call this hypothetical point o_1. The second candidate is the hypothetical point defined by the intersection of r_1, d_3, and c_2. Let us call this hypothetical point o_2.[6] Here, we reiterate that the preceding discussion holds for all values of odd $\eta \geq 3$.

---

[5] As expected r_1, c_1, and d_1 correspond to the row, column, and diagonal PCEs, respectively, of e_1.

[6] Hypothetical positions o_1 and o_2 are not shown in Figure III.7(a), since neither exists for this specific pattern.

From the preceding discussion, we note that it is not possible to have more than two additional *utips* apart from e_1. Next, we show that the two hypothetical points o_1 and o_2 may not exist simultaneously. Consider the 4-error pattern with error-positions defined by e_2, e_3, o_1, and o_2. We note that if o_1 and o_2 exist simultaneously, the following interactions must be satisfied:

$$o_1 \overset{r_1}{\sim} o_2, o_1 \overset{d_2}{\sim} e_2, o_1 \overset{c_3}{\sim} e_3 \qquad (\text{III.25})$$
$$o_2 \overset{d_3}{\sim} e_3, o_2 \overset{c_2}{\sim} e_2, e_2 \overset{r_2}{\sim} e_3.$$

Equation (III.25) implies that the 4-error pattern defined by positions e_2, e_3, o_1, and o_2 must constitute a valid RCD codeword, since the parities of each of the error positions are mutually satisfied. This, of course, is impossible, and we must conclude that the hypothetical points o_1 and o_2 cannot co-exist.

Of course, o_1 and o_2 cannot be the same point, since this would require two column PCEs, c_2 and c_3, to intersect (and two diagonal PCEs d_2 and d_3 to intersect). Thus, we can safely conclude that for a *class-2.1* 3-error pattern, at most one *utip* may exist other than e_1, and, further, that this *utip* must fall on the PCE of e_1 that is parallel to the PCE that contains e_2 and e_3.

Finally, we point out that specific cases exist where neither the PCEs r_1, c_2, and d_3, nor the PCEs r_1, c_3, and d_2, respectively, intersect at a common point. In this case we have no additional *utips* apart from e_1. The pattern in Figure III.7(a) is representative of such a case for $\eta = 5$. Also, such patterns can be easily generalized to odd $\eta \geq 3$. This completes the proof. $\square$

We note from the proof of Lemma III.6 that each of the $\eta - 1$ positions on r_1 other than e_1 uniquely determines a set of *class-2.1* 3-error patterns with exactly one additional *utip* apart from e_1. We label such *class-2.1* 3-error patterns as *category-1*. The proof of Lemma III.6 also suggests that we can follow a modified approach to exhaustively construct all possible *category-1 class-2.1* 3-error patterns. This result is presented in the following lemma.

Lemma III.7: *Consider a class-2.1 3-error pattern with error-positions e_1, e_2, and e_3, where the positions are labeled to satisfy $e_1 \nsim e_2$, $e_1 \nsim e_3$, and $e_2 \sim e_3$. There are exactly $3\eta^2 (\eta - 1) (\eta - 3)$ error-patterns in this class that have exactly one utip other than e_1.*

Proof: From Lemma III.6 we know that a *class-2.1* 3-error pattern can have either 0 or 1 additional *utip* apart from the 1-position e_1. We are interested in exhaustively constructing all *class-2.1* 3-error patterns that have exactly one additional *utip* apart from e_1. The construction steps are as follows:

1. Choose a position for e_1. There are η^2 different ways in which this can be done.

2. Choose a PCE type (either row, column, or diagonal) on which to locate e_2 and e_3. This can be achieved in $\binom{3}{2} = 3$ ways. Let us label this chosen PCE type as x.

3. Excluding position e_1, choose one of the other $\eta - 1$ positions, on the PCE of e_1 that is of type-x, as the desired *utip*. We label this point as t.

4. Consider the two PCEs other than of type-x passing through t. For convenience, we assume that x is of type *row*. We are then considering c_t and d_t as the column and diagonal PCEs passing through t.

5. If c_t and d_t, respectively, intersect any arbitrarily chosen row PCE r at positions that are both allowed for placing the two remaining 1-positions, then the two positions defined by the intersection of c_t and r, and d_t and r, are labeled as e_2 and e_3 (the order does not matter). Bit-positions e_1, e_2, and e_3 then constitute a *category-1 class-2.1* 3-error pattern.

Figure III.7(b) shows a specific example of a *category-1 class-2.1* 3-error pattern constructed according to this method for the RCD code with $\eta = 5$. It is important to remember that the described construction technique can be applied to all odd $\eta \geq 3$.

We now only need to determine the number of possible ways in which step (5) in the construction can be accomplished. Clearly, row r_1 is not a valid option to place e_2 and e_3. The intersection of c_t and d_1, which we label f_1, constitutes a forbidden position for e_2 and e_3, and, hence, the row corresponding to f_1 is also not a valid option to place e_2 and e_3 (refer Figure III.7(b) to better understand this constraint via a specific example). Similarly, the intersection of d_t and c_1, which we label f_2, represents a forbidden position for e_2 and e_3, and, hence, the row PCE of f_2 is also not a valid option to place e_2 and e_3.

Finally, we note that f_1 and f_2 cannot fall on the same row, as this would require e_1, t, f_1 and f_2 to constitute a valid codeword. Thus, we conclude that a total of three rows are forbidden for placing e_2 and e_3. Consequently, there are $(\eta - 3)$ different rows that can be chosen to place e_2 and e_3 once the *utip* has been determined. It follows that the number of different ways in which a *category-1 class-2.1* 3-error pattern can be constructed equals $3\eta^2 (\eta - 1) (\eta - 3)$. $\square$

We are now in a position to state and prove the final theorem of this subsection.

Theorem III.5: *For an RCD code with parameter η, at least $\frac{3}{2}\eta^2 (\eta - 1)(\eta - 3)$* *$(\eta - 4)$ class-2.1 3-error patterns decode successfully to the all-zeros codeword.*

Proof: From Lemma III.7 we know that exactly $3\eta^2 (\eta - 1) (\eta - 3)$ *class-2.1* 3-error patterns have a single *utip* apart from the 1-position e_1. Further, it was also shown in the proof of Lemma III.6 that for a *class-2.1* 3-error pattern, at most one additional *utip* can exist apart from the 1-position e_1. Thus, the number of *class-2.1* 3-error patterns with no additional *utip* apart from the bit-position e_1 is given by $|C2.1| - 3\eta^2 (\eta - 1) (\eta - 3)$, where $|C2.1| = \frac{3}{2}\eta^2 (\eta - 1) (\eta^2 - 5\eta + 6)$ is the total number of 3-error patterns belonging to *class-2.1* (refer equation (III.2)). Simplifying gives

$$|C2.1| - 3\eta^2 (\eta - 1) (\eta - 3) = \frac{3}{2}\eta^2 (\eta - 1) (\eta - 3) (\eta - 4) \qquad \text{(III.26)}$$

From Lemma III.5, it follows that all these *class-2.1* 3-error patterns must decode successfully to the all-zeros codeword. Since we know nothing conclusive about the successful decoding of the *category-1 class-2.1* 3-error patterns, the number $\frac{3}{2}\eta^2 (\eta - 1) (\eta - 3) (\eta - 4)$ must serve as a lower bound on the number of successful decoding to the all-zeros codeword. $\square$

6.   *Class-3* 3-error patterns and their decoding via BFA

In this subsection, we prove the following result that completely describes the decoding of *class-3* 3-error patterns via the BFA.

Theorem III.6: *For an RCD code with parameter η, each and every 3-error pattern belonging to class-3 always decodes successfully to the all-zeros codeword in exactly one iteration of the BFA.*

Proof: Consider a 3-error pattern belonging to *class-3*, and let the three bit-positions in error be labeled e_1, e_2, and e_3, respectively. All three bit-positions e_1, e_2, and e_3 participate in the same PCE, say h_1. Clearly, PCE h_1 will be unsatisfied. In addition, for each of e_1, e_2, and e_3, the remaining two PCEs in which each bit-position participates will also be unsatisfied. Thus, each of e_1, e_2, and e_3 has three unsatisfied PCEs associated with it.

Now consider any other bit-position, say o in the array. If o participates in h_1, then o will have only one unsatisfied PCE associated with it, since the orthogonality condition precludes o from participating with either of e_1, e_2, and e_3 again. If o does not participate in h_1, the structure of RCD codes ensures that o will always participate in a PCE, say h_2, that is in the same parallel bundle as h_1. By virtue of being in the same parallel bundle, h_1 and h_2 will not have any common bit-positions. Thus h_2 will not have any error positions participating in it, and will be satisfied. Consequently, if o does not participate in h_1, o can have at-the-most 2 unsatisfied PCEs associated with it. At the first BFA iteration, only the three bit positions in error, i.e., e_1, e_2, and e_3, will have the maximum number of three unsatisfied PCEs associated with each of them and will be flipped. Thus the *class-3* 3-error pattern will be corrected in one BFA iteration. □

7. *Class-2.3* 3-error patterns and their decoding via BFA

The final sub-class that must be tackled is *class-2.3*. To describe the BFA decoding of this class, we first prove the following lemma.

Lemma III.8: *For an RCD code with parameter η, η prime, the BFA decoding of each and every class-2.3 3-error pattern will result in an error-pattern of* *weight greater than three at the end of the first BFA iteration.*

Proof: Consider a 3-error pattern belonging to *class-2.3*, and let the three bit-positions in error be labeled e_1, e_2, and e_3, respectively. Without loss of generality, let us assume that e_1 and e_2 participate in a row PCE, say r_1. Then, e_1 and e_3 can participate in a column PCE or a diagonal PCE only. If we further assume that e_1 and e_3 participate in a column PCE, say c_1, then e_2 and e_3 must necessarily participate in a diagonal PCE, say d_1. Bit-position e_1 will have only its diagonal PCE, say d_2, unsatisfied. Similarly, bit-position e_2 will have only its column PCE, say c_2, unsatisfied, and bit-position e_3 will have only its row PCE, say r_2, unsatisfied. Thus, each of the three bit-positions in error in a *class-2.3* 3-error pattern has only one associated unsatisfied PCE.

Additionally, one will always have bit-positions that are not in error and that have at least two associated unsatisfied PCEs. An example would be the bit-position, say o_1 at the intersection of d_2 and c_2 (recall that two PCEs belonging to different parallel bundles must intersect at exactly one bit-position in the array). Similarly, the bit-position o_2 at the intersection of d_2 and r_2, and the bit-position o_3 at the intersection of c_2 and r_2, must also have at least two unsatisfied PCEs. Figure III.8 shows an example of such a such a situation for $\eta = 5$.

If we further utilize the fact that η is odd, we see that d_2, c_2, and r_2 cannot have a common intersection, since if they had a common intersection, say o, then the 4-error pattern with bit-positions e_1, e_2, e_3, and o would in fact constitute a codeword of Hamming weight 4 – an impossibility when η is odd. It follows that o_1, o_2, and o_3 are distinct bit-positions and each will have at least two associated unsatisfied PCEs. At the first BFA iteration, at least one of o_1, o_2, or o_3 will be flipped and, further, neither e_1, e_2, nor e_3 will be flipped. Thus, at the end of

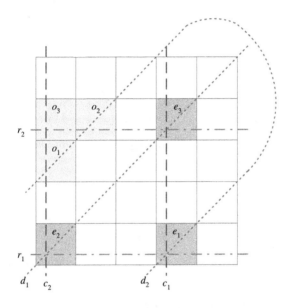

Figure III.8: Illustration for Lemma III.8, where e_1, e_2, and e_3 represent the three bit-positions in error, and o_1, o_2, and o_3 represent three other positions each having 2 unsatisfied PCEs.

the first BFA iteration, the resultant error-pattern will have weight greater than three. □

We have observed, based on an exhaustive enumeration and decoding of 3-error patterns, that for RCD codes all *class-2.3* 3-error patterns always result in decoder failure. While the preceding lemma highlights the difficulty associated with correctly decoding *class-2.3* 3-error patterns, it does not prove, by any means, that *class-2.3* 3-error patterns must always result in decoder failure. This result may be stated as a proposition as follows:

Proposition III.1: *For an RCD code with parameter η, η prime, each and every 3-error pattern belonging to class-2.3 results in decoder failure.*

We believe that this proposition can be proved by further studying the interactions between the error positions in the error pattern at the end of the first BFA iteration.

8. Interpreting the theoretical results

Until this point, we have presented numerous theorems that characterize the BFA decoding of 3-error patterns for RCD codes. Our ultimate goal is to use these results to construct a lower bound on the number of 3-error patterns that successfully decode to the all-zeros codeword, which in turn can be used to compute an upper and lower bound on the WER of RCD codes under BFA decoding.

From the results for *class-2.2* and *class-2.3*, we see that the possibility of any of such 3-error patterns decoding to the all-zeros codeword is indeed bleak. Thus, for these classes, we may lower-bound the number of error-patterns that decode successfully to the all-zeros codeword by 0. The penalty in using this lower-bound is not very severe if we also realize that *class-2.2* and *class-2.3* have negligible contribution to the total number of 3-error patterns, especially as η increases[7].

From Theorem III.6 we know that all $|C3| = \frac{\eta^2(3\eta-3)(\eta-2)}{3!}$ *class-3* 3-error patterns decode successfully to the all-zeros codeword. Further, from Theorem III.5 at least $\frac{3}{2}\eta^2(\eta-1)(\eta-3)(\eta-4)$ *class-2.1* decode successfully to the all-zeros codeword. Lastly, from Theorem III.3 and the discussion on the BFA decoding of *class-1* 3-error patterns, we have a strong argument to support the conjecture that at least $|C1| - 2A_6$ *class-1* 3-error patterns will successfully decode to the all-zeros codeword, where $|C1| = \frac{\eta^2(\eta^2-3\eta+2)(\eta^2-6\eta+10)}{3!}$ and $A_6 = \eta\binom{\eta}{3}$ is the number of codewords at the minimum distance of 6. Combining the previous results, we can claim that at least

$$z_3 = |C3| + \frac{3}{2}\eta^2(\eta-1)(\eta-3)(\eta-4) + |C1| - 2A_6 \qquad \text{(III.27)}$$

3-error patterns will decode successfully to the all zeros codeword.

We can use the value of z_3 to construct a theoretical upper bound on the WER of arbitrary RCD codes. Further, assuming that all the remaining $\binom{\eta^2}{3} - z_3$ 3-error patterns lead to either decoder failure or decoding to wrong codewords, we can also construct a theoretical lower bound on WER of RCD codes. Both these tasks will be discussed in detail in Section III.I, where we present plots of various performance curves for three different RCD codes. Before proceeding to the plots, in the next two sections we provide some experimental results based on the exhaustive generation, classification, and decoding of 3-error and 4-error patterns for RCD codes for a range of values of η.

---

[7]Exhaustive enumeration results for 3-error patterns that are presented in Section III.F indicate that this lower-bound is, in fact, exact.

F. Experimental Results for 3-Error Patterns and Some Propositions

We now present experimental results based on an exhaustive enumeration, class-ification, and BFA decoding of all 3-error patterns for values of η for which such analysis is feasible. The number of 3-error patterns increases with increasing η, and the increase is approximately proportional to η^6.

Beginning with a particular value of η, the following tasks are performed in the experiment:

1. The parity-check matrix of the RCD code is first generated.

2. All $\binom{\eta^2}{3}$ possible 3-error patterns are sequentially generated according to lexicographical order [11].

3. Each 3-error pattern is classified based on the five classes and sub-classes defined in Subsection III.E.1, and the counter for that class of error patterns is incremented.

4. The 3-error pattern, then, is decoded according to the BFA with maximum number of iterations set to 20 (20 is sufficiently high for the case of 3-error patterns). Decoder success or failure, the number of iterations to decode, and the Hamming weight of the output of the decoder are noted, and corresponding counters for the 3-error pattern classes are incremented.

5. For each class, numbers for the cardinality of the class, error patterns that decoded successfully to the all-zeros codeword, error patterns that resulted in decoder failure, and error patterns that decoded to codewords other than

the all-zeros codeword are obtained via the experiments. For each class, histograms of the number of iterations to decode and the Hamming weights of the decoder output are also generated.

Table III.1 lists the results of the exhaustive enumeration of 3-error patterns for odd η beginning with $\eta = 5$ and going up to $\eta = 23$. For each class of 3-error patterns, the total number of error patterns, the number that decode successfully to the all-zeros codeword, the number that result in decoder failure, and the number that decode to codewords other than the all-zeros codeword, are listed. We also plot, in Figures III.9, III.10, and III.11, the bar graphs of the data from the columns of Table III.1 corresponding to $\eta = 7$, $\eta = 17$, and $\eta = 23$, respectively. The increasing significance of *class-1* and *class-2.1* 3-error patterns with increasing η is clearly observed. Further, we also observe that in both these classes, the fraction of error-patterns that successfully decode to the all-zeros codeword increases (and approaches unity) with increasing η.

Table III.1 contains a wealth of interesting behavior and patterns. Additionally, the results of Table III.1 agree perfectly with the theorems proven in Section III.E, and in many cases indicate stronger behavior than what has already been proven. The most significant fact that is verified by the data in Table III.1 is that the theoretical lower bound on the number of 3-error patterns given by equation (III.27) is, in fact, exact for prime η. For non-prime η, the number of *class-1* 3-error patterns that successfully decode to the all-zeros codeword is seen to be marginally higher than $|C1| - 2A_6$. Based on the values of η considered in Table III.1, the principal observations in addition to those already proven are:

1. All the 3-error patterns belonging to *class-2.3* result in decoder failure.

Table III.1

Experimental results for 3-error patterns for $\eta = 5, 7, 9, \ldots, 23$.

Class	statistics	5	7	9	11	13	15	17	19	21	23
1	total	250	4165	27972	117975	375518	989625	2277320	4731627	9077250	16333933
	success	150	3675	26487	114345	368082	976050	2254200	4694805	9021537	16252467
	fail	0	490	1458	3630	7436	12600	23120	36822	55566	81466
	incorr.	100	0	27	0	0	975	0	0	147	0
2.1	total	900	8820	40824	130680	334620	737100	1456560	2651184	4524660	7331940
	success	300	5292	29160	101640	273780	623700	1262352	2339280	4048380	6633660
	fail	600	3528	11664	29040	60840	113400	194208	311904	476280	698280
	incorr.	0	0	0	0	0	0	0	0	0	0
2.2	total	900	4410	13608	32670	66924	122850	208080	331398	502740	733194
	success	0	0	0	0	0	0	0	0	0	0
	fail	600	3528	11664	29040	60840	113400	194280	311904	476280	698280
	incorr.	300	882	1944	3630	6084	9450	13872	19494	26420	34914
2.3	total	100	294	648	1210	2028	3150	4624	6498	8820	11638
	success	0	0	0	0	0	0	0	0	0	0
	fail	100	294	648	1210	2028	3150	4624	6498	8820	11638
	incorr.	0	0	0	0	0	0	0	0	0	0
3	total	150	735	2268	5445	11154	20475	34680	55233	83790	122199
	success	150	735	2268	5445	11154	20475	34680	55233	83790	122199
	fail	0	0	0	0	0	0	0	0	0	0
	incorr.	0	0	0	0	0	0	0	0	0	0

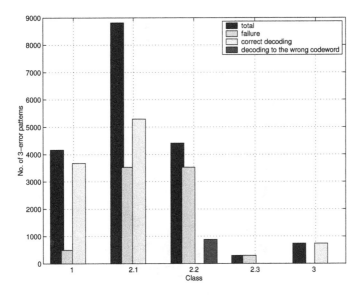

Figure III.9: Results of exhaustive search for 3-error patterns for RCD code with $\eta = 7$.

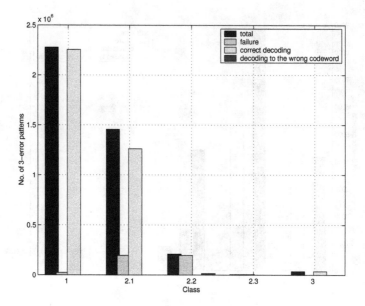

Figure III.10: Results of exhaustive search for 3-error patterns for RCD code with $\eta = 17$.

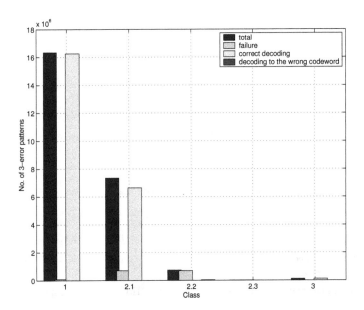

Figure III.11: Results of exhaustive search for 3-error patterns for RCD code with $\eta = 23$.

2. None of the 3-error patterns of *class-2.2* decode successfully to the all-zeros codeword. They lead to either decoder failure or incorrect decoding.

3. *class-2.1* 3-error patterns either decode successfully to the all-zeros codeword or lead to decoder failure. In other words, none of the 3-error patterns of *class-2.1* get decoded incorrectly (i.e., to codewords other than the all-zeros codeword)

4. The number of error patterns that result in decoder failure for *class-2.1* and *class-2.2* are equal.

5. For prime $\eta > 5$, none of the 3-error patterns of *class-1* get decoded incorrectly. On the other hand, for non-prime η, there always exist *class-1* 3-error patterns that get decoded incorrectly.

6. For prime η, the number of 3-error patterns of *class-1* that decode successfully can be exactly given by the expression $\eta^2 \left(\eta^2 - 3\eta + 2 \right) \left(\eta^2 - 6\eta + 8 \right) / 3!$, which is also equal to $|C1| - 2A_6$, where $A_6 = \eta^2 \left(\eta^2 - 3\eta + 2 \right) / 3!$ is the number of codewords with Hamming weight equal to $d_{min} = 6$ [55].

The lemmas and theorems presented in Section III.E provide a stepping stone to proving the stronger results, which are suggested by the data presented in Table III.1, and are enumerated in the preceding list.

G. Classification of 4-Error Patterns

The methodology developed in the previous section to classify 3-error patterns can be readily extended to classify and study error-patterns at higher Hamming weights. In this section we consider error-patterns at Hamming weight 4.

Let e_1, e_2, e_3, and e_4 represent the four bit-positions in error. Clearly, since $d_{\min} = 6 > 4$, at least one PCE that is not satisfied will always exist for 4-error patterns. Similar to 3-error patterns (see Section III.E), the primary classification of 4-error patterns is performed on the basis of the maximum number of error-positions that participate in a single PCE. Four different major classes are possible in the case of 4-error patterns for RCD SPC codes for $\eta \geq 5$. These are as follows:

Class-1:   None of the four bit-positions in error participate with each other in a PCE. In other words, there is at most only one error-position participating in any PCE. Based on the notation we introduced earlier, membership to *class-1* is defined by the set of bit-positions that satisfy

$$e_1 \nsim e_2, e_1 \nsim e_3, e_1 \nsim e_4$$
$$e_2 \nsim e_3, e_2 \nsim e_4, e_3 \nsim e_4 \ . \qquad \text{(III.28)}$$

Class-2:   There is at least one PCE in which two error-positions participate, and there is no PCE that has three or more of the error-positions jointly participating.

Class-3:   There is at least one PCE in which three error-positions participate, and there is no PCE that has all four of the error-positions jointly participating.

Class-4:   All four bit-positions in error participate in the same PCE. In other words, there is at least one PCE in which all the four error-positions jointly participate. The orthogonality condition ensures that there is exactly one such PCE in the case of RCD SPC codes. Further, for a 4-error pattern of this class, no PCE can have exactly two or exactly three error-positions jointly participating (due to the orthogonality condition).

In the major classification presented above, there is no further sub-classification possible in *classes-1* and *4*. We observe, however, that *class-2* can be further sub-classified based on how many PCEs have exactly two participating error-positions.

1. Subclassification of *class-2* 4-error patterns

For regular LDPC codes, six possibilities exist for *class-2* since there are $\binom{4}{2} = 6$ ways in which the four error-positions can be paired, and it is possible to have anywhere between any one of the six pairs to all six of the six pairs simultaneously participating in distinct PCEs. Note that the orthogonality condition ensures that any pair can jointly participate in at most one PCE. Thus, the following sub-classification arises in *class-2*:

Class-2.1: Exactly one PCE has a pair of participating error-positions, and all other PCEs have at most one participating error-position. Such a case arises, for example, when $e_1 \sim e_2$ but $e_1 \nsim e_3$, $e_1 \nsim e_4$, $e_2 \nsim e_3$, $e_2 \nsim e_4$, $e_3 \nsim e_4$.

Class-2.2: Exactly two distinct PCEs each have a pair of error-positions participating, and every other PCE has at most one participating error-position.

Class-2.3:   Exactly three distinct PCEs each have a pair of participating error-positions, and every other PCE has at most one participating error-position.

Class-2.4:   Exactly four distinct PCEs each have a pair of participating error-positions, and every other PCE has at most one participating error-position.

Class-2.5:   Exactly five distinct PCEs each have a pair of participating error-positions, and every other PCE has at most one participating error-position.

Class-2.6:   Exactly six distinct PCEs each have a pair of participating error-positions, and every other PCE has at most one participating error-position. By the orthogonality condition, each of a the six PCEs that have a pair of participating error-positions must have a distinct pair.

We notice that a further level of subclassification is possible for *classes-2.2, -2.3,* and *-2.4.* The defining characteristics of these subclassifications are described by means of representative 4-error patterns. First consider the next level of sub-classification for *class-2.2,* which is as follows:

Class-2.2.0:   The two PCEs, with each having a pair of participating error-positions, do not have a common bit position. For example, $e_1 \overset{p_1}{\sim} e_2$, $e_3 \overset{p_2}{\sim} e_4$, and no other pairs jointly participate in PCEs, where p_1 and p_2 represent two distinct PCEs.

Class-2.2.1: The two PCEs, with each having a pair of participating error-positions, have a common bit position. For example, $e_1 \overset{p_1}{\sim} e_2$, $e_2 \overset{p_2}{\sim} e_3$, and no other pairs jointly participate in PCEs.

Of course, thanks to the orthogonality condition, the two PCEs of a *class-2.2* error-pattern that each have two participating error-positions cannot have both the error-positions common and, hence, we see that all possibilities under *class-2.2* are hereby exhausted.

Next, we consider the subclassification for *class-2.3*:

Class-2.3.1: The three PCEs, with each having a pair of participating error-positions, have the same common bit position. For example, $e_1 \overset{p_1}{\sim} e_2$, $e_1 \overset{p_2}{\sim} e_3$ $e_1 \overset{p_3}{\sim} e_4$, and no other pairs jointly participate in PCEs. Here, again, p_1, p_2 and p_3 represent the three distinct PCEs. We note that common error-positions among PCEs each having a pair of participating error-positions are unavoidable when there are more than two such PCEs.

Class-2.3.2: Two of the three PCEs, with each having a pair of participating error-positions, have a common error-position (say e_1). Similarly, two other PCEs of the three have a common error-position (say e_2 that is different from e_1). Thus, an example would be $e_1 \overset{p_1}{\sim} e_2$, $e_1 \overset{p_2}{\sim} e_3$ $e_2 \overset{p_3}{\sim} e_4$.

Class-2.3.3: Each pair of PCEs of the three PCEs, with each having exactly two of participating error-positions, has a distinct common error-position. An example of this category would be $e_1 \overset{p_1}{\sim} e_2$, $e_1 \overset{p_2}{\sim} e_3$ $e_2 \overset{p_3}{\sim} e_3$. This is the only subclassification for *class-2.3* that does not have each of the four error-positions participating pairwise in at least one PCE (in the provided example e_4 does not participate in p_1, p_2, or p_3).

We have exhausted, therefore, all possible subclasses for *class-2.3*.

Finally, the subclassification for *class-2.4* is:

Class-2.4.1: One of the four error-positions participates in only one of the four PCEs that have exactly two participating error-positions. A representative is $e_1 \overset{p_1}{\sim} e_2$, $e_1 \overset{p_2}{\sim} e_3$, $e_2 \overset{p_3}{\sim} e_3$, and $e_1 \overset{p_4}{\sim} e_4$ (observe that e_4 appears in only one of the four PCEs p_1, p_2, p_3, p_4).

Class-2.4.2: Each of the four error-positions participates in exactly two of the four PCEs that have two participating error-positions. An example is $e_1 \overset{p_1}{\sim} e_2$, $e_1 \overset{p_2}{\sim} e_3$, $e_2 \overset{p_3}{\sim} e_4$, and $e_3 \overset{p_4}{\sim} e_4$.

Again, bearing in mind that the orthogonality condition must be satisfied by the PCEs of the RCD code, we see that any viable *class-2.4* 4-error pattern must belong to either of the two subclasses presented above.

2. Subclassification of *class-3* 4-error patterns

Similar to *class-2*, subclassification is also possible in *class-3*. It is based on the number of error-positions that participate with the lone error-position not in the PCE with three participating error-positions, and is as follows:

Class-3.0: The single error-position, which does not participate in the PCE with three participating error-positions, does not participate via other PCEs with any of the three other positions.

This class can be represented as $e_1 \overset{p_1}{\sim} e_2 \overset{p_1}{\sim} e_3$, and $e_4 \nsim e_1$, $e_4 \nsim e_2$, and $e_4 \nsim e_3$.

Class-3.1: The single error-position, which does not participate in the PCE with three participating error-positions, participates pairwise with only one of the three other positions via a different PCE. This class can be represented as $e_1 \overset{p_1}{\sim} e_2 \overset{p_1}{\sim} e_3$, and $e_4 \overset{p_2}{\sim} e_1$, but $e_4 \nsim e_2$, and $e_4 \nsim e_3$.

Class-3.2: The single error-position, which does not participate in the PCE with three participating error-positions, participates pairwise with two of the three other positions via two different PCEs. This class can be represented as $e_1 \overset{p_1}{\sim} e_2 \overset{p_1}{\sim} e_3$, $e_4 \overset{p_2}{\sim} e_1$, and $e_4 \overset{p_3}{\sim} e_2$, but $e_4 \nsim e_3$.

Class-3.3: The single error-position, which does not participate in the PCE with three participating error-positions, participates pairwise with each of the three other error-positions via three different PCEs. This class can be represented as $e_1 \overset{p_1}{\sim} e_2 \overset{p_1}{\sim} e_3$, $e_4 \overset{p_2}{\sim} e_1$, $e_4 \overset{p_3}{\sim} e_2$, and $e_4 \overset{p_4}{\sim} e_3$.

Again, by construction of the subclasses, we are assured of exhausting all 4-error patterns of *class-3*. This completes the classification structure for 4-error patterns.

3. Exhaustiveness of the classification scheme and related combinatorial expressions

A few remarks are now in order. First, we can show that for RCD codes, 4-error patterns of *class-2.6* and *class-3.3* cannot be realized. The proofs for these results are presented next.

Lemma III.9: *For RCD codes, there cannot exist any 4-error pattern that belongs to class-2.6.*

Proof: Let us assume that a 4-error pattern belonging to *class-2.6* exists. Note that there must be exactly six distinct PCEs, each having exactly two participating error-positions. Since we have four error-positions in all, we can choose a pair of error-positions in $\binom{4}{2} = 6$ different ways. Thus each possible pair of error-positions must participate pairwise in a distinct PCE. Further, each error-position has only three distinct PCEs. It follows that, for every error-position in a *class-2.6* 4-error pattern, all its PCEs are satisfied (sum to 0 modulo 2). In other words, the *class-2.6* error pattern with four error-positions must constitute a valid codeword. This is known, however, to be an impossibility because $d_{\min} = 6$, and leads to a contradiction. $\square$

Lemma III.10: *For RCD codes, there cannot exist any 4-error pattern that belongs to class-3.3.*

Proof: Let us assume that a pattern belonging to *class-3.3* exists. Without loss of generality, we can define this pattern by the following interactions:

$$e_1 \overset{r_1}{\sim} e_2 \overset{r_1}{\sim} e_3,$$

$$e_4 \overset{p_1}{\sim} e_1, e_4 \overset{p_2}{\sim} e_2, e_4 \overset{p_3}{\sim} e_3 , \tag{III.29}$$

where r_1 represents a row PCE, and p_1, p_2, and p_3 represent three other different and distinct PCEs. Note that e_4 cannot participate in PCE r_1 (if not, then our error pattern will belong to *class-4* rather than *class-3.3*). Since e_4 participates in all three PCEs p_1, p_2, and p_3, it follows that one of p_1, p_2, or p_3 must be a row PCE. Again, without loss of generality, we can assume that p_1 is a row PCE. Thus, we require two distinct row PCEs r_1 and p_1 to intersect at e_1, which is possible only if $r_1 = p_1$. This, of course, violates our assumption, and thus, a 4-error pattern that belongs to *class-3.3* cannot exist. □

The second remark concerns the exhaustiveness of the suggested classification scheme. It is clear that the major classification (based on the maximum number of error-positions participating in a PCE) is exhaustive, and no RCD code 4-error pattern can fall outside these four major classes. We have also given convincing arguments in favor of the exhaustiveness of the sub-classification schemes for *class-2* and *class-3*. We can safely conclude, consequently, that the classification scheme described in the preceding discussion exhaustively partitions the set of all 4-error patterns.

Finally, with a little bit of effort, combinatorial expressions for the number of patterns belonging to each class/sub-class can be computed. These are presented in Table III.2. Here, the notation $|Cx|$ is used to indicate the cardinality of *class-x*. Expressions are derived constructively for all subclasses except *class-1*. The exhaustiveness of our classification scheme must then be invoked to compute $|C1|$, the number belonging to *class-1*. Thus, if $|C2|$ and $|C3|$ represent the sum of all 4-error patterns belonging to all possible subclasses of *class-2* and *class-3*, respectively, and $|C4|$ represents the number belonging to *class-4*, then we can write

$$|C1| = \binom{\eta^2}{4} - (|C2| + |C3| + |C4|) \ , \qquad (\text{III}.30)$$

where $\binom{\eta^2}{4}$ is the total number of 4-error patterns possible.

We also note that the presented expressions agree exactly with experimental results obtained via exhaustive generation and classification of 4-error patterns for RCD codes with $\eta \in \{5, 7, \ldots, 17\}$ (see Table III.3). Also, for the indicated values of η, these experiments did not result in any 4-error pattern falling outside the constructed classification scheme, thus supporting the exhaustiveness assertion for the classification scheme.

The classification of 4-error patterns presented so far demonstrates that we have to contend with 16 different possibilities, while only 5 different classification possibilities exist for the 3-error patterns. Thus, we see a clear indication that as the Hamming weight of the error-pattern is increased beyond 4, the number of classification possibilities increases rapidly, making it extremely difficult to continue a combinatorial analysis of error-patterns.

Thankfully, however, since the decoding errors resulting from 3- and 4-error patterns dominate the word-error rate at high signal-to-noise ratios, pursuing the analysis of w-error patterns for $w \geq 5$ is not really necessary.

H.   Experimental Results for 4-Error Patterns and Some Observations

Similar to the 3-error patterns case, we now present experimental results based on an exhaustive enumeration, classification, and BFA-decoding of all 4-error

Table III.2

Combinatorial expressions for the cardinality of the different classes/subclasses of 4-error patterns.

Class	Number of constituent 4-error patterns
$|C2.1|$	$3\eta\binom{\eta}{2}\{\left(\frac{(\eta-2)(\eta-3)}{2}\right) - \binom{\eta-3}{2}(3\eta-7)\}$
$|C2.2.0|$	$3\eta\binom{\eta}{2}\binom{\eta-3}{2}(3\eta-7)$
$|C2.2.1|$	$3\eta^2(\eta-1)(\eta-3)^3$
$|C2.3.1|$	$\eta^2(\eta-1)\left((\eta-3)^2 + (\eta-2)\right)$
$|C2.3.2|$	$6\eta\binom{\eta}{2}(\eta-3)(2\eta-5)$
$|C2.3.3|$	$2\eta\binom{\eta}{2}(\eta-3)^2$
$|C2.4.1|$	$3\eta^2(\eta-1)(\eta-3)$
$|C2.4.2|$	$\frac{3}{2}\eta\binom{\eta}{2}(3\eta-7)$
$|C2.5|$	$3\eta\binom{\eta}{2}$
$|C2.6|$	0 (see lemma III.9)
$|C3.0|$	$3\eta\binom{\eta}{3}(\eta-3)(\eta-4)$
$|C3.1|$	$18\eta\binom{\eta}{3}(\eta-3)$
$|C3.2|$	$18\eta\binom{\eta}{3}$
$|C3.3|$	0 (see lemma III.10)
$|C4|$	$3\eta\binom{\eta}{4}$
$|C1|$	Can be computed as $\binom{\eta^2}{4} - (|C2| + |C3| + |C4|)$

Table III.3

Experimental results for 4-error patterns for $\eta = 5, 7, \ldots, 17$.

Class				η				
statistics		5	7	9	11	13	15	17
1	total	75	8869	160866	1334630	7078565	28080675	90848884
	success	0	0	22113	542685	4276545	20557800	73545876
	fail	0	3577	88695	530585	1968512	5433300	12544912
	incorr.	75	5292	50058	261360	833508	2089575	4758096
2.1	total	1050	46746	545292	3317820	13856310	45274950	124438776
	success	0	441	50301	600765	3644316	15214500	49245600
	fail	600	27342	338256	2079990	8389836	25415100	65739408
	incorr.	450	18963	156735	637065	1822158	4645350	9453768
2.2.0	total	600	18522	145800	660660	2190240	5925150	13885872
	success	0	0	2916	7260	42588	210600	762960
	fail	150	12348	62694	337590	1195506	3271050	7698960
	incorr.	450	6174	80190	315810	952146	2443500	5423952
2.2.1	total	2400	56448	419904	1858560	6084000	16329600	38064768
	success	0	0	0	0	0	0	0
	fail	1800	35280	271188	1255980	4356144	12220200	29519616
	incorr.	600	21168	148716	602580	1727856	4109400	8545152
2.3.1	total	700	6174	27864	88330	225108	494550	975664
	success	0	0	0	0	0	0	0
	fail	0	1176	5832	4840	32448	39600	184960
	incorr.	700	4998	22032	83490	192660	454950	790704
2.3.2	total	3000	31752	151632	493680	1277640	2835000	5632032
	success	0	0	0	0	0	16200	27744
	fail	1200	22932	87480	297660	888264	1701000	3468000
	incorr.	1800	8820	64152	196020	389373	1117800	2136288
2.3.3	total	400	4704	23328	77440	202800	453600	906304
	success	0	0	0	0	0	0	0
	fail	400	2940	14580	55660	154128	356400	739840
	incorr.	0	1764	8748	21780	48672	97200	166464
2.4.1	total	600	3528	11664	29040	60840	113400	194208
	success	0	0	0	0	0	0	0
	fail	600	3528	11664	29040	60840	113400	194208
	incorr.	0	0	0	0	0	0	0

Table III.3
Continued

Class statistics		η						
		5	7	9	11	13	15	17
2.4.2	total	600	3087	9720	23595	48672	89775	152592
	success	0	0	0	0	0	0	0
	fail	0	0	0	0	0	0	0
	incorr.	600	3087	9720	23595	48672	89775	152592
2.5	total	150	441	972	1815	3042	4725	6936
	success	0	0	0	0	0	0	0
	fail	0	0	0	0	0	0	0
	incorr.	150	441	972	1815	3042	4725	6936
2.6	total	0	0	0	0	0	0	0
	success	0	0	0	0	0	0	0
	fail	0	0	0	0	0	0	0
	incorr.	0	0	0	0	0	0	0
3.0	total	300	8820	68040	304920	1003860	2702700	6311760
	success	300	8820	68040	304920	1003860	2702700	6311760
	fail	0	0	0	0	0	0	0
	incorr.	0	0	0	0	0	0	0
3.1	total	1800	17640	81648	261360	669240	1474200	2913120
	success	0	5292	37908	152460	438048	1036800	2164032
	fail	1200	10584	40824	101640	219024	421200	721344
	incorr.	600	1764	2916	7260	12168	16200	27744
3.2	total	900	4410	13608	32670	66924	122850	208080
	success	0	0	0	0	0	0	0
	fail	300	2646	10692	25410	54756	106650	180336
	incorr.	600	1764	2916	7260	12168	16200	27744
3.3	total	0	0	0	0	0	0	0
	success	0	0	0	0	0	0	0
	fail	0	0	0	0	0	0	0
	incorr.	0	0	0	0	0	0	0
4	total	75	735	3402	10890	27885	61425	121380
	success	75	0	486	0	2028	12825	27744
	fail	0	735	729	1815	3042	2700	6936
	incorr.	0	0	2187	9075	22815	45900	86700

patterns for values of η for which such analysis is feasible.[8]

The tasks performed in the experiment are similar to those for 3-error patterns (see Section III.F). We limit ourselves to a maximum of 20 BFA-iterations. Table III.3 lists the results of the exhaustive enumeration of 4-error patterns for $\eta \in \{5, 7, \ldots, 17\}$. For each class of 4-error patterns, we list the total number of error patterns, the number that decode successfully to the all-zeros codeword, the number that result in decoder failure, and the number that decode to codewords other than the all-zeros codeword. We point out that the total numbers for each class/sub-class obtained by exhaustive experiments for all considered values of η agree exactly with the total number computed via the corresponding class/sub-class combinatorial expression from Table III.2. The results presented in Table III.3 can be used to compute upper and lower bounds on the WER of RCD codes for the considered values of η. This concept is discussed in detail in Section III.I.

Further, it is also possible to extract some general trends and behavior patterns from the data to assist us in making predictions for larger values of η. We, however, do not proceed along this path. Rather, we compare the performance bounds based on the decoding of only 3-error patterns and the performance bounds based on the decoding of 3- and 4-error patterns to demonstrate that the bounds based on decoding of only 3-error patterns are extremely tight at high SNR and, hence, sufficient to characterize RCD code performance at these

---

[8]The number of 4-error patterns increases rapidly with increasing η, When $\eta = 17$, we must generate about 2.85×10^8 4-error patterns, which is close to the limit of the number of error-patterns that can be analyzed in reasonable time (1 week to 1 month) on a Pentium-4 2GHz processor.

SNR values. The additional information provided by knowledge of decoding of 4-error patterns serves to improve the tightness of the WER bounds only over a very small range of SNR values. While this is not insignificant, this range of SNR values corresponds to moderately low WER error rates that are capable of being estimated fairly easily using standard Monte Carlo techniques for a wide range of RCD codes. As such, the extremely complicated combinatorial analysis of the decoding of 4-error patterns is not necessary to obtain good WER estimates for RCD codes in a reasonable amount of time. The one area where the combinatorial analysis of 4-error patterns may prove beneficial is in the performance evaluation of RCD codes on the binary symmetric channel with erasures (BSC/E). The reader is referred to Chapter IV of this dissertation for an in-depth discussion on RCD code performance on the BSC/E.

I. Performance Bounds for RCD Codes on the BSC Based on Decoding of 3- and 4-Error Patterns

Consider a binary symmetric channel (BSC) with transition probability p. For a code of length n, the input to the bounded-distance decoder (BDD), or for that matter any hard-decision decoder, is a binary vector of the same length that we call the *received word*.

Three different mutually exclusive and exhaustive outcomes are possible at the output of a general hard-decision decoder (this includes the BDD). They are:

1. The received word gets correctly decoded to the codeword that was trans-

mitted. The associated probability, P_C, is the probability of correct decoding.

2. The received word gets decoded to a codeword other than the one that was transmitted, i.e., a wrong codeword. The associated probability, P_{DWC}, is the probability of decoding to a wrong codeword.

3. The decoder is unable to decode the received word. In this case the decoder signals a failure. The associated probability, P_F, is the probability of failure.

Clearly, $P_C + P_{DWC} + P_F = 1$ and, thus, the *word error rate* (WER) is given by $WER = 1 - P_C = P_{DWC} + P_F$, which is the probability of not decoding to the correct codeword and comprises two components.

Now consider specifically the case of a BDD. For a BDD with error correcting capability $t = \lfloor \frac{d_{min}-1}{2} \rfloor$ [75],

$$P_C = \sum_{i=0}^{\lfloor \frac{d_{min}-1}{2} \rfloor} \binom{n}{i} p^i (1-p)^{n-i} \qquad \text{(III.31)}$$

and

$$WER_{BDD} = 1 - P_C = 1 - \sum_{i=0}^{\lfloor \frac{d_{min}-1}{2} \rfloor} \binom{n}{i} p^i (1-p)^{n-i}. \qquad \text{(III.32)}$$

Further,

$$P_{DWC} = \sum_{j=d_{min}}^{n} A_j \sum_{k=0}^{\lfloor \frac{d_{min}-1}{2} \rfloor} P_k^j, \qquad \text{(III.33)}$$

where

$$P_k^j = \sum_{r=0}^{k} \binom{j}{k-r} \binom{n-j}{r} p^{j-k+2r} (1-p)^{n-j+k-2r}, \qquad \text{(III.34)}$$

and A_j represents the number of codewords at Hamming weight j. The set $\mathcal{A} = \{A_0, A_1, \ldots, A_n\}$ is called the weight enumerator (WE) of the code.

At low values of p, P_{DWC} is dominated by the first term in the sum of equation (III.33), i.e., $A_{d_{\min}} P_k^{d_{\min}}$. Thus P_{DWC} can be bounded as

$$P_{DWC} \geq A_{d_{\min}} P_k^{d_{\min}}, \qquad \text{(III.35)}$$

where the bound is tight for low values of p.

The BSC can be derived from a binary phase-shift-keying (BPSK) additive white Gaussian noise (AWGN) model with signaling at $\pm V$, variance σ^2, and a single decision threshold at 0. In such a case $p = Q(V/\sigma)$, where $Q(x) = \frac{1}{\sqrt{2\pi}} \int_x^\infty e^{\frac{-t^2}{2}} dt$.

Given that $d_{\min}$, $A_{d_{\min}}$, and the code rate $R = \frac{(\eta-1)(\eta-2)}{\eta^2}$ for the RCD codes are known, we can plot WER based on equation (III.32), and the bound on P_{DWC} based on equation (III.33), vs. p or vs. $\frac{E_b}{N_0} = \frac{V^2}{2R\sigma^2}$ for different values of η. Such curves are commonly called performance curves. Also, the difference between the value of WER and P_{DWC} at any value of $\frac{E_b}{N_0}$ equals the probability of failure P_F of the BDD. The performance curves based on BDD serve as bounds on the BFA decoder performance given that the BFA is at least a BDD for the RCD code (see Theorem III.2).

BFA decoding of RCD codes, however, is capable of correcting a large number of error-patterns of weight ≥ 3. The theoretical lower-bound obtained on the number of 3-error patterns that successfully decode to the all-zeros codeword (see equation (III.27)) can be used to further strengthen the bounds on the WER of the BFA. In particular, for BFA decoding of RCD codes,

$$P_{C,\text{BFA}} \geq \left(\sum_{i=0}^{2} \binom{n}{i} p^i (1-p)^{n-i} \right) + \frac{z_3}{\binom{n}{3}} p^3 (1-p)^{n-3} \qquad \text{(III.36)}$$

and, hence,

$$\text{WER}_{\text{BFA}} = 1 - P_{C,\text{BFA}} \leq 1 - \left(\sum_{i=0}^{2} \binom{n}{i} p^i (1-p)^{n-i} \right) - \frac{z_3}{\binom{n}{3}} p^3 (1-p)^{n-3} ,$$
$$\text{(III.37)}$$

where z_3 is computed via equation (III.27). This newly obtained bound helps us accurately predict the performance of the BFA for RCD codes at high values of E_b/N_0 (i.e., low WER, P_{DWC}, and P_F) without resorting to time-and-resource-consuming Monte Carlo simulations to obtain performance curves.

The theoretical upper-bound on the WER of BFA given by equation (III.37) can also be compared with the upper-bound on WER of BFA based on our experimental results of the exhaustive enumeration and decoding of 3-error patterns for those RCD codes for which such data is available. Thus, for those η for which 3-error pattern experimental results exist, the WER of the BFA can be upper-bounded by

$$1 - \sum_{i=0}^{3} c_i \binom{n}{i} p^i (1-p)^{n-i}, \qquad \text{(III.38)}$$

where c_i is the fraction of i-error patterns that the BFA successfully decodes to the all-zeros codeword. Clearly, $c_i = 1$ for $i = 0, 1, 2$, and c_3 is obtained from Table III.1.

As observed previously in Section III.F, for RCD codes with prime η, the theoretical lower bound on the number of 3-error patterns given by equation (III.27) is, in fact, exact for prime η. Hence, the upper-bound on WER computed by equations (III.37) and (III.38) will be identical. Further, given this observation of exactness, we may also construct a theoretical lower bound on the WER of the BFA for RCD codes as

$$\text{WER}_{\text{BFA}} \geq \left(1 - \frac{z_3}{\binom{n}{3}} \right) p^3 (1 - p)^{n-3} \, , \tag{III.39}$$

since we know that exactly $\left(1 - \frac{z_3}{\binom{n}{3}} \right)$ 3-error patterns lead to either decoder failure or DWC, and there are many more error-patterns at higher weights that lead to decoder failure or DWC.

Similarly, based on only the experimental results, the WER of the BFA may be lower-bounded by

$$\left(1 - \frac{c_3}{\binom{n}{3}} \right) p^3 (1 - p)^{n-3}, \tag{III.40}$$

where c_3 is obtained from Table III.1.

Finally, along similar lines, if experimental results from the exhaustive evaluation of 4-error patterns exists for an RCD code with a specific η, then one may upper bound the WER of the BFA by

$$1 - \sum_{i=0}^{4} c_i \binom{n}{i} p^i (1 - p)^{n-i}, \tag{III.41}$$

where $c_i = 1$ for $i = 0, 1, 2$, and c_3 and c_4 are obtained from Table III.1 and Table III.3, respectively. Similarly, one can also lower bound the WER of the BFA by

$$\sum_{i=3}^{4} \left(1 - \frac{c_i}{\binom{n}{i}}\right) p^i \left(1 - p\right)^{n-i}, \qquad \text{(III.42)}$$

where c_3 and c_4 are again obtained from Table III.1 and Table III.3, respectively.

Figures III.12, III.13, and III.14 show the performance curves for RCD codes with $\eta = 17$, $\eta = 23$, and $\eta = 37$, respectively, based on the various bounds discussed previously.

In all figures, the dotted curve represents the WER of the BDD computed via equation (III.32). The circles represent the lower-bound on P_{DWC} of the BDD based on equation (III.35). The $\times$ symbol represents the BFA WER obtained via standard Monte Carlo simulations with a maximum of 20 iterations. The dashed line gives the absolute lower bound on WER computed based on the Shannon limit for the BSC [3] (called the Shannon lower bound). Note that at moderate-to-high E_b/N_0, the standard Monte Carlo results are significantly lower than the WER given by equation (III.32), reaffirming the observation that the BDD performance of RCD codes is a poor approximation of the BFA performance.

We also have plotted the 99% confidence intervals of the standard Monte Carlo estimates based on the Gaussian assumption [36] as vertical bars about the standard Monte Carlo WER. The verticals bars may not be clearly discernible at high WER ($\gtrsim 10^{-4}$) due to the high confidence of the Monte Carlo estimates in this region.

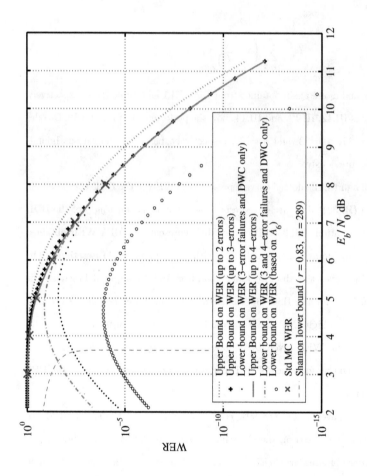

Figure III.12: Performance curves for the RCD code with $\eta = 17$ for BDD on the BSC (rate = 0.83).

114

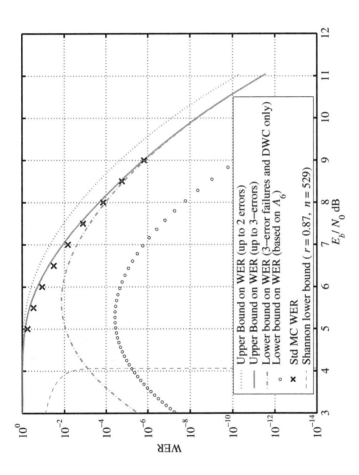

Figure III.13: Performance curves for the RCD code with $\eta = 23$ for BDD on the BSC (rate $= 0.87$).

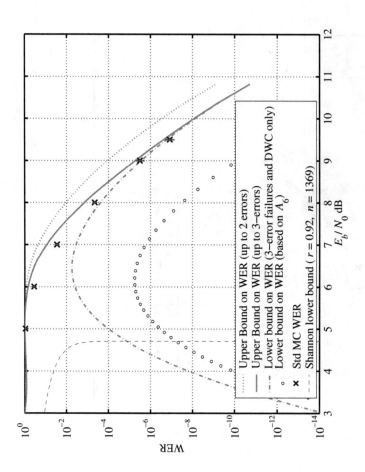

Figure III.14: Performance curves for the RCD code with $\eta = 37$ for BDD on the BSC (rate = 0.92).

In Figure III.12, the + symbols represent the upper bound on the WER of the BFA computed via equation (III.37) (which is identical to the bound based on experimental results computed via equation (III.38)), while the dots represent the theoretical lower bound on the WER of the BFA obtained via equation (III.39) (which, in turn, is identical to the bound based on experimental results computed via equation (III.40)). Observe that these two curves merge together for $E_b/N_0 \geq 8$dB (or for WER of $\leq 10^{-4}$). Thus, for higher E_b/N_0 (or lower WER), the decoder WER performance predicted by these two bounds can be deemed to be highly accurate. The fact that the standard MC results fall between the two curves further substantiates the accuracy of these bounds.

For the case of $\eta = 17$ we also have experimental results based on the exhaustive enumeration and decoding of 4-error patterns (see Table III.3). The solid curve denotes the upper-bound on BFA WER computed via equation (III.41), while the dash-dot curve corresponds to the lower-bound on the BFA WER computed via equation (III.42). These two curves merge together for $E_b/N_0 \geq 6.5$dB (or for WER of $\leq 10^{-2}$). From Figure III.12, we notice that the additional information from considering both 3- and 4-error patterns over considering only 3-error patterns helps us improve the bounds on the WER only over a small range in the moderate SNR region. While this is certainly useful, it is certainly not necessary if one needs to predict decoder performance at very low WER (or high SNR), since for this purpose the curves based on the + symbols and dots suffice. Of course, the standard MC results also fall within the solid curve and the dash-dot curve.

In Figure III.13, the solid line represents the upper bound on the WER of the BFA computed via equation (III.37) (which is identical to the bound based on experimental results computed via equation (III.38)), while the dash-dot line represents the theoretical lower bound on the WER of the BFA obtained via equation (III.39) (which, in turn, is identical to the bound based on experimental results computed via equation (III.40)). For the $\eta = 23$ RCD code, these two curves merge together for $E_b/N_0 \geq 9$dB (or for WER of $\leq 10^{-6}$). Thus for higher E_b/N_0 (or lower WER), the decoder WER performance predicted by these two bounds can be deemed to be highly accurate. The standard MC results again fall between the two curves further substantiating the accuracy of these bounds. For the $\eta = 23$ RCD code, since we do not have experimental results based on the exhaustive enumeration and decoding of 4-error patterns, the corresponding bounds defined by equations (III.41) and (III.42) cannot be plotted.

Finally, the description of the various performance curves for the $\eta = 37$ RCD code in Figure III.14 is identical to those for the $\eta = 23$ RCD code in Figure III.13. Here the solid curve and the dash-dot curves represents purely theoretical bounds (via equations (III.37) and (III.39), respectively) since exhaustive enumeration results do not exist for 3-error patterns (and needless to say also for 4-error patterns). Here the solid and dash-dot curves merge together for $E_b/N_0 \geq 9.5$dB (or for WER of $\leq 10^{-7}$) and, thus, the decoder WER performance predicted by these two bounds can be deemed to be highly accurate for higher E_b/N_0 (or lower WER). An interesting thing to note in Figure III.14 is that the standard Monte Carlo result at 9.5 dB falls below the lower bound (dash-dot curve). While this may seem to invalidate our theoretical bounds, that is not necessarily true since the standard MC WER estimate at 9.5dB is based on only 25 error events and, hence, has relatively low reliability. In particular, the upper end of the 99% confidence interval goes well above the lower bound (dash-dot curve).

From the performance plots for the three RCD codes (Figures III.12, III.13, and III.14), we see that the theoretical results of BFA decoding of 3-error patterns enables us to improve the upper bound on the WER of BFA decoding over the upper-bound based on BDD only (dotted curve). Further, this improved upper-bound is also tight at high SNR.

J. Conclusions

1. Summary and conclusions

In this paper we have analyzed the decoding of error patterns for the RCD code via the bit-flipping algorithm (BFA). We showed that the BFA is capable of correcting all error patterns up to Hamming weight equal to $\lfloor \frac{d_{min}-1}{2} \rfloor = 2$, where $d_{min} = 6$ is the minimum distance of the RCD code. Thus, for the RCD code, we showed that the BFA is capable of performing at least as well as a bounded-distance decoder.

We then analyzed the decoding of errors patterns of Hamming weight-3 via the BFA. A classification scheme was developed for the 3-error patterns based on the interactions, via parity-check equations (PCEs), of the three positions in error. Combinatorial expressions were obtained for the number of error patterns belonging to each of the classes. Important theoretical results characterizing the decoding under BFA of the various classes and sub-classes of 3-error patterns were proved. These theoretical results were then used to construct a lower bound on the number of 3-error patterns that decode successfully to the all-zeros codeword.

We then presented experimental results for the exhaustive generation, classification, and BFA decoding of 3-error patterns for RCD codes with $5 \leq \eta \leq 23$. It was seen that the obtained results agreed perfectly with the derived analytical results. Additionally, analysis of the experimental results also indicated the possibility of existence of stronger results that characterize the BFA decoding of error patterns of the various classes of weight-3.

Similarly, a classification scheme was also provided for 4-error patterns, and experimental results based on the exhaustive generation, classification, and BFA decoding of 4-error patterns for RCD codes with $5 \leq \eta \leq 17$ were also obtained.

Upper and lower bounds on the word-error rate (WER) performance of certain RCD codes on a binary symmetric channel were computed, based on the theoretical results on the number of 3-error patterns that decode successfully to the all-zeros codeword. WER upper- and lower-bounds from exhaustive searches of 3-error patterns only, as well as of 3- and 4-error patterns, were also computed for those values of η where such computations were possible. For those ηs where results from an exhaustive search were available, the bounds from the exhaustive search were compared with the theoretical upper and lower WER bounds. The theoretical lower and upper bounds based on 3-error patterns were identical to the corresponding experimental exhaustive search results for 3-error patterns only. All the computed bounds were also compared with standard Monte Carlo simulations results of WER and were found to agree well.

We conclude that the classification scheme for 3- and 4-error patterns presented in this chapter is exhaustive. However, extending this classification to higher-weight error patterns results in a rapid increase in the complexity of the process due to the progressive larger number of classes and sub-classes. We also conclude that a thorough combinatorial analysis of 3-error patterns under BFA decoding provides us with important theoretical results to construct tight bounds on RCD code WER performance.

From the performance bounds, one can accurately characterize the WER performance of RCD codes at high signal-to-noise ratios (SNRs). Also, the theoretical/combinatorial analysis of the decoding of various classes of 3-error patterns provides us with a means to compute these bounds without resorting to time-consuming simulations for the entire class of RCD codes.

Further, while analysis of error-patterns of weight > 3 can help in providing us with tighter performance bounds over a larger range of SNR, we conclude from our observations and results that the study of decoding of 3-error patterns for the class of RCD codes is sufficient to characterize the code performance at high SNR, and it is possible to obtain these performance curves via a combinatorial analysis of the decoding of 3-error patterns.

Finally, the analysis presented in this chapter based on the decoding of error-patterns of specific Hamming weight has led to the development of an importance sampling [2] technique that is effective in estimating a code's hard-decision decoding performance under certain conditions that may be easily achieved for a range of code parameters. This technique, which we call the SNR-invariant importance sampling (IIS) technique, is presented in Chapter VI.

2. Future research

The material presented in this chapter raises a plethora of interesting possibilities, which may be pursued for future research. These research possibilities are as follows:

1) The proof of the stronger results on the decoding of various classes of 3-error patterns suggested by the analysis of data in Table III.1 and presented in Section III.F constitutes interesting research, since these results would help us ascertain the exact number of instances of decoder failures and incorrect decoding for each class or subclass of 3-error patterns.

2) One also may perform a thorough analysis of the BFA decoding of 4-error patterns and, in particular, that of the asymptotically dominant classes, to further strengthen some of the analytical bounds.

3) The research presented in this chapter can be used hopefully to obtain bounds on the BFA decoding performance for the general class of codes based on diagonals in square-arrays. In particular, it would be interesting to prove or disprove that the BFA is capable of successfully decoding all error patterns up to weight $\gamma - 1$ for a square array code constructed with γ different slope sets defining the parity-check matrix.

4) The number of decoding failures observed under BFA decoding of 3-error patterns far exceeds the minimum number of 3-error patterns that one would expect to lead to decoder failure (This number equals $\binom{6}{3}A_6$ and corresponds to all those 3-error patterns that are equidistant from the all-zeros codeword and a codeword at Hamming weight 6). This suggests the potential for improving the BFA. One possibility that immediately comes to mind is the use of a randomized rule to flip only one of all those error positions that have the maximum number of unsatisfied PCEs associated with them. Other modifications could also be attempted.

5) A combinatorial analysis of error-patterns is infeasible for codes in general,

122

unless there is an underlying structure in the code that is simple enough to be tractable under such analysis. Another class of codes that may lend itself to combinatorial analysis of error patterns is the class of n-dimensional single parity-check product (SPCP) codes [12]. We recommend performing such a combinatorial analysis to further extend the technique proposed in this chapter and to evaluate the performance bounds for the class of n-dimensional SPCP codes.

Chapter IV.

ON MINIMUM-WER
PERFORMANCE OF FEC CODES
FOR THE BSC/E BASED ON
BPSK-AWGN AND EXTENDED
HARD-DECISION DECODING

In this chapter, we study the word-error-rate (WER) performance and the WER-minimizing decision thresholds t^* for forward error correction (FEC) codes on the binary symmetric channel with erasures (BSC/E) with decoding via an extension to the bounded-distance decoder (BDD) for the binary symmetric channel (BSC). We also study the performance of RCD codes under decoding via the extension to the BFA (e-BFA) by deriving some bounds and evaluating performance via standard Monte Carlo simulations. We compare the performance of the e-BFA with decoding via a message passing decoder designed for the BSC/E. As we indicated previously in this dissertation, the BSC/E represents the simplest channel that provides soft-information at the decoder and the desire to study code performance on the BSC/E is motivated by the possibility that a substantial performance improvement may be possible by just using a BSC/E as compared the the BSC/E.

A. Introduction

1. The BSC/E

The binary symmetric channel with erasures (BSC/E) is a discrete-input discrete-output channel [75] with input alphabet $\{-1, 1\}$ and output alphabet $\{-1, E, 1\}$, where E denotes an erasure. Let x denote the input to the BSC/E and y the output. The transition probabilities $\alpha = Pr\{y = E | x = -1\} = Pr\{y = E | x = 1\}$ and $\beta = Pr\{y = 1 | x = -1\} = Pr\{y = -1 | x = 1\}$ with $0 \leq \alpha,\ \beta,\ (1 - \alpha - \beta) \leq 1$ completely describe the BSC/E. The binary symmetric channel (BSC) is a special case of the BSC/E with $\alpha = 0$.

The BSC/E can be derived from a BPSK-AWGN model with signaling at ± 1, AWGN variance σ^2, and a pair of decision thresholds at $\pm t$, $t \geq 0$. The BSC/E and its BPSK-AWGN generating model is shown in Figure IV.1. The output y of the decision circuit is given as

$$ y = \begin{cases} -1 & r < -t \\ E & -t \leq r \leq t \\ 1 & r > t\,, \end{cases} \qquad \text{(IV.1)} $$

where r is the received value at the AWGN channel output. The parameters α and β of the BSC/E can then be computed as $\alpha = Q\left(\frac{-t+1}{\sigma}\right) - Q\left(\frac{t+1}{\sigma}\right)$ and $\beta = Q\left(\frac{t+1}{\sigma}\right)$, where $Q(x) = \frac{1}{\sqrt{2\pi}} \int_x^\infty e^{\frac{-t^2}{2}} dt$.

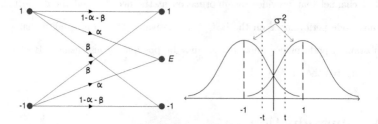

Figure IV.1: The BSC/E and its BPSK-AWGN generating model

It should be noted that not all possible combinations of values of α and β can be generated for the BSC/E using the BPSK-AWGN system. Fortunately, this is not a very serious limitation of the system since, in this study, we are particularly interested in a BSC/E that arises from three-level decisioning (i.e., with two thresholds) at the receiver of a communication system with binary transmission and AWGN. Further, setting $t = 0$ results in a BSC (since $\alpha = 0$) with $\beta = Q\left(\frac{1}{\sigma}\right)$.

2. Motivation to study the BSC/E

FEC codes used on communication systems with a single-threshold decision rule at the receiver are usually decoded based on hard-decision decoding algorithms, which take binary values for each of the codeword bits as inputs. Such a hard-decision rule destroys all the information that may be gleaned from the knowledge of the exact value of the received signal. It is common knowledge that the performance of FEC codes can be improved by the availability of soft-information at the output of the decision device. Theoretically, soft-information is equivalent to infinite precision on the received signal. In practice, soft-information implies having anywhere from 3 to 5 bits of information on the received signal, i.e., from 2^3 to 2^5 decision levels at the detector. The utilization of soft-information to decode FEC codes presents a trade-off between FEC code performance on the one hand, and decoder and receiver implementation speed and complexity on the other.

The BSC/E with two thresholds and three decision levels represents the simplest example of a channel model with soft-information at its output. The two-threshold decision circuit of the BSC/E can be considered as a parallel arrangement of two one-threshold detectors. It is thus much simpler than a 3-bit or 5-bit A/D converter and is more likely to be capable of implementation at ultra high speeds. Further, as we will explain in the next section, decoding of FEC codes for the BSC/E can be carried out by a parallel utilization of two hard-decision decoders, followed by some elementary decisioning. Such a decoder for the BSC/E has computational complexity that is slightly greater than twice the complexity of the hard-decision decoder employed. Finally, it has also been shown [62, 65] that for certain codes such as low-density parity-check (LDPC) codes, a major portion of the coding gain that can be achieved in going from a BSC with hard-decision iterative decoding to a BPSK-AWGN channel model with soft-decision iterative decoding occurs in simply going to the BSC/E. The BSC/E channel model, thus, seems like a promising alternative to obtain improved communication system performance, without an excessive increase in receiver and decoder complexity.

B. FEC Decoding for the BSC/E and Characterization of the Word Error Rate

The decoding of binary forward error correction (FEC) codes for the BSC/E can be achieved by a simple extension of any hard-decision FEC decoder for the BSC (henceforth, referred to as the constituent decoder (CD)) as given in [75]. In the sequel, we will refer to this decoding algorithm for the BSC/E simply by using the qualifiers *extended* or *extension of*. The CD operates on binary variables with values of -1 or 1, which correspond to the logic values 1 and 0, respectively. Given a received word, $\mathbf{w}$, at the output of the BSC/E, we decode according to the following:

1. Replace every location in $\mathbf{w}$ that has an erasure with 1 and decode using the CD to obtain a resulting codeword $\mathbf{c}_1$.

2. Then replace every location in $\mathbf{w}$ that has an erasure with a -1 and decode using the CD to obtain a resulting codeword $\mathbf{c}_{-1}$.

3. Compute the Hamming distance between the pair $\mathbf{w}$ and $\mathbf{c}_1$ and the pair $\mathbf{w}$ and $\mathbf{c}_{-1}$. An erasure location in $\mathbf{w}$ will always have a unit contribution to both these Hamming distances since neither $\mathbf{c}_1$ nor $\mathbf{c}_{-1}$ has any erasure locations.

4. Finally, the CD codeword (i.e., either $\mathbf{c}_1$ or $\mathbf{c}_{-1}$) that has the lesser Hamming distance from the received word $\mathbf{w}$ is chosen as the BSC/E decoder output. If the Hamming distances between $\mathbf{w}$ and $\mathbf{c}_{-1}$, and $\mathbf{w}$ and $\mathbf{c}_1$ are equal, and $\mathbf{c}_{-1} \neq \mathbf{c}_1$, then a decoder failure is signaled[1], while if $\mathbf{c}_{-1} = \mathbf{c}_1$, then either one can be chosen as the BSC/E decoder output.

Consider a code with length n and minimum distance $d_{\min}$. A bounded-distance decoder (BDD) for such a code for the BSC, by definition is capable of correcting up to e errors in a received word as long as $e \leq \lfloor \frac{d_{\min}-1}{2} \rfloor$ [75]. If the CD is a BDD for the BSC, then it can be shown that the extension of the BDD for the BSC/E (or, equivalently, the extended BDD) can correct all patterns of e errors and f erasures as long as $(2e + f) < d_{\min}$ [75]. Note that steps 1 to 3 can be performed in parallel with two separate constituent decoders to increase decoding speed.

The word error rate (WER) for an FEC decoder can be defined as the sum of the probability of decoding to a wrong codeword (P_{DWC}) and the probability of decoder failure (P_F) (or equivalently, as $1 - P_C$, where P_C is the probability of correct decoding). For the extended BDD, the WER can be computed as

$$\text{WER} = 1 - \sum_{\substack{e,f \geq 0; \\ (2e+f) < d_{\min}}} \binom{n}{f} \binom{n-f}{e} \alpha^f \beta^e (1 - \alpha - \beta)^{n-e-f} , \qquad \text{(IV.2)}$$

where e and f are the number of errors and erasures, respectively.

Equation (IV.2) can be generalized for the extension of any hard-decision decoder to the BSC/E via

$$\text{WER} = 1 - \sum_{\substack{e,f \geq 0; \\ (e+f) \leq n}} \gamma_{e,f} \binom{n}{f} \binom{n-f}{e} \alpha^f \beta^e (1 - \alpha - \beta)^{n-e-f} , \qquad \text{(IV.3)}$$

---

[1]One may alternatively flip a fair coin and choose either $\mathbf{c}_1$ or $\mathbf{c}_{-1}$ as the BSC/E decoder output in this situation.

where $\gamma_{e,f}$ denotes the fraction of patterns with e errors and f erasures that can be decoded successfully to the transmitted codeword. For the extended BDD, we have $\gamma_{e,f} = 1$ if $(2e+f) < d_{min}$ and $\gamma_{e,f} = 0$ otherwise. Thus, equation (IV.3) simplifies to equation (IV.2) for the extended BDD. Note that the introduction of the fractions $\gamma_{e,f}$ automatically accounts for constituent decoders that are capable of correcting error-patterns with Hamming weight $> \lfloor \frac{d_{min}-1}{2} \rfloor$.

The $\gamma_{e,f}$ are dependent on the chosen code and CD. In some special cases (such as when CD=BDD), it is rather straightforward to compute $\gamma_{e,f}$. In general, however, the $\gamma_{e,f}$ terms must be determined by theoretical analysis of the decoder for the code under consideration, and/or by experimental enumeration and decoding of error-erasure patterns.

C. WER Minimizing Thresholds

A hard-decision decoder for the BSC estimates the transmitted codeword based on the binary received word, and this estimated codeword depends completely on the binary received word. In particular, given the binary received word, the decoder estimate is independent of the value of the transition probabilities that define the BSC. A similar claim can be made for the extension of a hard-decision decoder to the BSC/E – that the codeword estimated by the decoder depends completely on the received word with ternary symbols. From the preceding argument, it follows that specifying a code and CD completely determines the fractions $\gamma_{e,f}$ in equation (IV.3), and that these fractions are independent of the parameters t and σ^2 defining the BSC/E transition probabilities. Thus, α and β are the only quantities in each term on the right hand side of equation (IV.3) that depend on the parameters t and σ^2. The parameter σ^2, which defines the AWGN, can be assumed to be beyond the control of the receiver for all practical purposes. On the other hand, the BSC/E decision threshold t is completely controlled at the receiver.

For a specific code and CD, and a particular value of σ^2, one can attempt to minimize equation (IV.3) with respect to $t \geq 0$. We first would be required to show the existence of a minimum, and if the minimum exists, we could then attempt to analytically describe the threshold t^* that achieves the minimum WER.

The partial derivative of α and β with respect to t can be easily computed as

$$\frac{\partial}{\partial t}\beta = \frac{-1}{\sqrt{2\pi\sigma^2}}e^{-\frac{(t+1)^2}{2\sigma^2}} \tag{IV.4}$$

$$\frac{\partial}{\partial t}\alpha = \frac{-1}{\sqrt{2\pi\sigma^2}}e^{-\frac{(-t+1)^2}{2\sigma^2}} + \frac{1}{\sqrt{2\pi\sigma^2}}e^{-\frac{(t+1)^2}{2\sigma^2}} . \tag{IV.5}$$

For the sake of argument, suppose we assume the existence of a minimum for equation (IV.3) as a function of t: we observe that employing equations (IV.4) and (IV.5) to analytically evaluate the expression for t^* satisfying $\frac{\partial}{\partial t}\text{WER}(t) = 0$ is extremely complicated and mathematically intractable. Hence, we resort to numerical minimization of WER based on equation (IV.3) with respect to t.

Exact knowledge of the terms $\gamma_{e,f}$ is necessary for numerical minimization. As discussed previously, this information, while known for the extended BDD, is not generally available for extension of other CDs. Consequently, in the remainder of this paper, we primarily focus our attention on the BDD extended to the BSC/E. Our objective is to study the behavior of t^* as a function of the channel signal-to-noise ratio (CSNR $\triangleq 10 \log_{10} \left(\frac{1}{2\sigma^2} \right)$) for different combinations of code parameters n and $d_{\min}$. The definition of CSNR naturally follows from the assumption of BPSK signaling a ± 1 and an AWGN channel with one-sided noise power spectral density $N_0 = 2\sigma^2$.

A further indication that a zero-derivative point for equation (IV.2) may not exist is provided by the following observation. Consider a hypothetical code of length n and $d_{\min} = 1$. For such a code, equation (IV.2) simplifies to

$$\text{WER} = 1 - (1 - \alpha - \beta)^n . \tag{IV.6}$$

For the BSC/E based on BPSK-AWGN,

$$\alpha + \beta \geq p , \tag{IV.7}$$

where $p \triangleq Q \left(\frac{1}{\sigma} \right)$ denotes the transition probability of the BSC based on the same BPSK-AWGN model. Combining equation (IV.6) and (IV.7), we get the result

$$\text{WER}(0) \leq \text{WER}(t) \tag{IV.8}$$

for all $t \geq 0$, where the WER is represented as a function of t for a fixed value of σ^2. From equation (IV.6), we also see that the WER is a monotonically increasing function of $t \geq 0$. Clearly, a zero-derivative point for equation (IV.6) does not exist in such a case. The above example also demonstrates that in the absence of an FEC code (represented by the situation where $n = 1$, and $d_{\min} = 1$), employing a BSC/E channel model instead of a BSC channel model always results in a degradation of the error rate for the same value of AWGN variance.

D. Procedure, Plots, Observations, and Discussion

In this section we study the behavior of the WER as a function of the dual-threshold parameter t for different CSNR, and the WER-minimizing threshold and the minimum WER as a function of CSNR for the BDD extended to the BSC/E. Given values of code parameters n and $d_{\min}$, we choose a CSNR and evaluate the WER via equation (IV.2) for a range of finely spaced values of t. We assign the value of t that minimizes the WER for that CSNR as t^*. The WER corresponding to t^* is WER(t^*). The above process is repeated for a range of values of CSNR, and t^* and WER(t^*) are plotted vs. CSNR. We also plot the quantity $\log_{10}$ WER$(t = t^*) - \log_{10}$ WER$(t = 0)$ versus the CSNR. The quantity $\log_{10}$ WER$(t = t^*) - \log_{10}$ WER$(t = 0)$ represents the *maximum improvement* in WER possible (in terms of orders of magnitude) by employing two decision thresholds at the receiver and decoding via the extended BDD over using a single decision threshold at the receiver and decoding via the BDD.

We perform two different sets of studies. In the first set, we fix $d_{\min} = 6$ and generate the WER vs. t plots, and t^*, WER(t^*), and $\log_{10}$ WER$(t^*) - \log_{10}$ WER(0) vs. CSNR plots, for values of $n \in \{49, 289, 529, 1369\}$. These values of n are specifically chosen since they correspond to the length of various RCD codes ($\eta = 7, 17, 23, 37$, respectively), where the class of RCD array codes have $d_{\min} = 6$. Since the code rates for the RCD array codes parameterized by η are known to be $R_c = \frac{(\eta-1)(\eta-2)}{\eta^2}$, we also plot WER$(t^*)$ and $\log_{10}$ WER$(t^*) - \log_{10}$ WER(0) vs. $E_b/N_0 \triangleq 10 \log_{10}\left(\frac{1}{2R_c\sigma^2}\right)$. In the second set, we fix $n = 529$ and generate the various plots for values of $d_{\min} \in \{6, 8, 9, 16, 17\}$. Both odd and even values of d_{min} are considered, where the codes with even $d_{min} > 6$ are obtained by progressively imposing additional slope diagonal parity-check equations on the $n = 529$ RCD code constructed on the 23×23 square array, while the odd d_{min} codes are obtained by discarding the overall parity bit from appropriate even d_{min} codes. Note that discarding the overall parity bit reduces the code length to 528. However, we continue using a code length of 529 in the analysis for convenience. The difference in performance introduced by this approximation is negligible. Since it is known that the amount of improvement possible in extending the BDD from the BSC to the BSC/E is significantly larger for even d_{min} than for odd d_{min} [51], this second set study will allow us to quantify the amount of improvement under our code assumptions.

1.   Fixed $d_{min} = 6$, different code lengths

Figure IV.2 shows the WER vs. t plot for a code with parameters $n = 49$ and $d_{min} = 6$. These parameters correspond to the RCD code with $\eta = 7$. Each thick solid curve corresponds to a specific CSNR value as indicated. It is clear, especially from the curves corresponding to high CSNR values, that as t is increased from 0 the WER initially decreases until it reaches a minimum at t^*, after which the WER increases. The dash-dot line with circles in Figure IV.2 connects the t^* values corresponding to the different CSNRs. At a given CSNR, the separation between the WER achieved at $t = 0$ (corresponding to a BSC) and the WER achieved at t^* indicates the maximum improvement possible (over employing a single decision threshold), by employing two decision thresholds at the receiver and decoding via the extended BDD. For example, from Figure IV.2 we observe that at CSNR = 12 dB, a maximum improvement of about six orders of magnitude is achieved, from $\approx 10^{-20}$ when $t = 0$ to $\approx 10^{-26}$ when $t = t^* = 0.166$. Equivalently, at WER $= 10^{-15}$, there is a coding gain of ≈ 1.0dB in employing the optimal BSC/E as opposed to the BSC.

Similarly, Figures IV.3, IV.4, and IV.5 show the WER vs. t plot for the $n = 289$, $d_{\min} = 6$ code, the $n = 529$, $d_{\min} = 6$ code, and the $n = 1369$, $d_{\min} = 6$ code. Plots for all the four cases considered exhibit similar overall behavior for the range of CSNR values plotted, with t^* seen to progressively increase with increasing CSNR and decreasing WER. Note that this non-decreasing behavior of t^* with respect to CSNR is not observed at extremely low values of CSNR (< -2 dB) as we will see later. We intentionally use the CSNR here since it does not depend on the code rate. We reiterate, however, that these parameters are chosen specifically because they correspond to RCD codes. For any specific CSNR and t, the shorter length code in this set of $d_{\min} = 6$ codes achieves a lower WER, which is perfectly reasonable to expect since the resulting BSC/E transition probabilities are the same for the codes in this set, but the same CSNR $= R_c E_b / N_0$ value will correspond to a lower E_b / N_0 for the longer code due to R_c increasing with n.

Figure IV.6 compares plots of the WER-minimizing threshold t^* vs. CSNR for codes with $d_{\min} = 6$ but different n. We see that for a given CSNR, t^* decreases with n, and the behavior of t^* vs. CSNR is similar for all n considered. As CSNR decreases from a high value (say 15 dB), t^* decreases monotonically until CSNR ≈ 0 dB. As CSNR decreases further, t^* stabilizes at a particular value; if the CSNR decreases even further, t^* begins to increase. Some of the other important observations associated with Figure IV.6 are as follows:

1. t^* decreases with n at all CSNR values considered.

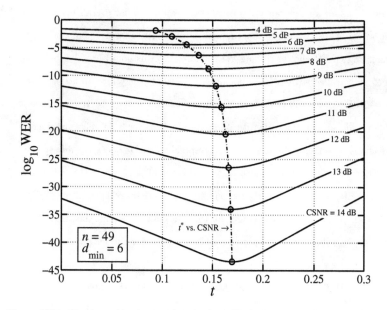

Figure IV.2: For $(n = 49,\ d_{\min} = 6)$ code: $\log_{10}$WER vs. t vs. CSNR and t^* vs. CSNR.

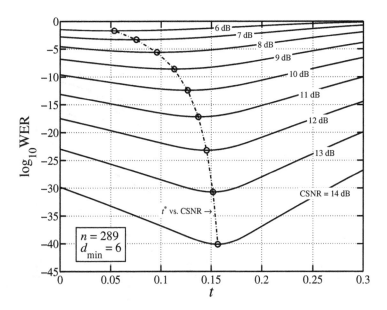

Figure IV.3: For $(n = 289,\ d_{\min} = 6)$ code: $\log_{10}$WER vs. t vs. CSNR and t^* vs. CSNR.

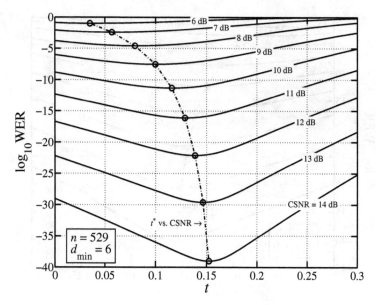

Figure IV.4: For $(n = 529, d_{\min} = 6)$ code: $\log_{10}$WER vs. t vs. CSNR and t^* vs. CSNR.

138

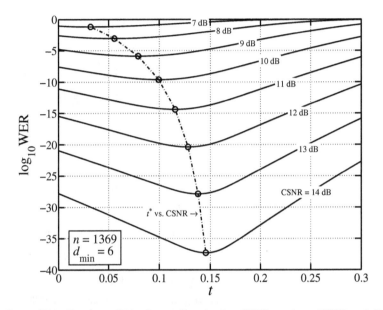

Figure IV.5: For $(n = 1369,\ d_{\min} = 6)$ code: $\log_{10}$WER vs. t vs. CSNR and t^* vs. CSNR.

2. For each n, the t^* vs. CSNR curve has a minimum $(t^*_{\min})$ at some CSNR < 0dB. Further, the range of CSNR values where $t^* \approx t^*_{\min}$ broadens as n increases.

3. The sharpest increase in t^* occurs for $0 \leq$ CSNR ≤ 12dB, with the location shifting to the larger values of CSNR as n increases.

Figure IV.7 compares plots of WER(t^*) and $\log_{10}$ WER$(t^*) - \log_{10}$ WER(0) vs. CSNR for codes with $d_{\min} = 6$ but different n. From Figure IV.7 we see that for a given CSNR, the shorter code achieves a lower value of WER(t^*). Note that for CSNR < 2 dB, the WER is very close to unity for all code lengths considered, and corresponds to extremely poor channel conditions. We also see that for any given CSNR, the shorter code has larger performance improvement. For example, at a CSNR of 10 dB, the $n = 49$ code has $t^* = 0.1588$ and WER$(t^*) = 2.1 \times 10^{-16}$, while the WER obtained by setting $t = 0$ is WER$(0) = 1.15 \times 10^{-12}$, giving a performance improvement of about 3.74 orders of magnitude. Corresponding values at the same CSNR, for all the code lengths considered, are presented in Table IV.1(a), and clearly show that the $n = 49$ code has the largest performance improvement of the lot. Equivalently, the coding gain at WER $= 10^{-15}$ is $\approx$ 1.0dB. Table IV.2(a) shows the coding gain in dB when the optimal BSC/E is employed instead of the BSC for the codes with $d_{\min} = 6$. Note that the coding gain in dB is the same whether it is considered in terms of CSNR or E_b/N_0.

In Figure IV.8, we plot WER(t^*) and $\log_{10}$ WER$(t^*) - \log_{10}$ WER(0) vs. E_b/N_0, where the code rates for computing the E_b/N_0 values are indicated next to the length of the code. The $n = 49$ code has the best WER(t^*) performance up to $E_b/N_0 = 9$ dB, beyond which the longer codes give the best performance. The curves corresponding to the three longer (and higher rate) codes are relatively

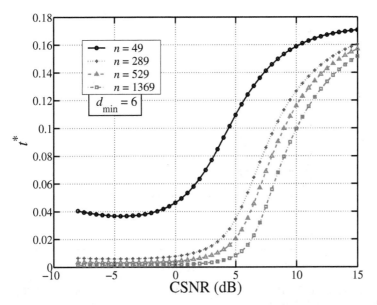

Figure IV.6: t^* vs. CSNR for codes with $d_{\min} = 6$ but different n.

141

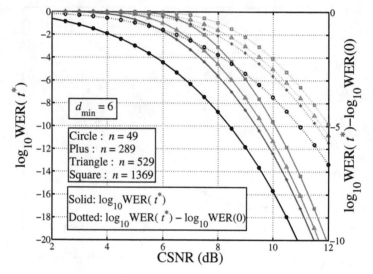

Figure IV.7: WER(t^*) vs. CSNR, and WER improvement between BSC/E ($t = t^*$) and BSC ($t = 0$) vs. CSNR, for codes with $d_{\min} = 6$ but different n.

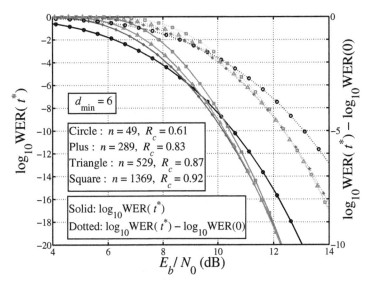

Figure IV.8: WER(t^*) vs. E_b/N_0, and WER improvement between BSC/E ($t = t^*$) and BSC ($t = 0$) vs. E_b/N_0, for codes with $d_{\min} = 6$ but different n and R_c.

very close to each other for almost the entire range of E_b/N_0 values plotted. Two things can be inferred from Figure IV.8. First, nearly similar performance for a variety of high rate (> 0.8) RCD codes can be achieved by employing the respective WER-minimizing decision thresholds t^* for a range of E_b/N_0 (9dB onwards in Figure IV.8), thus providing a trade-off between the choice of code-lengths and code-rate. Second, at any value of E_b/N_0, one can determine an optimal value of n (and hence R_c) for the class of RCD codes that provides the best WER performance with the decision threshold chosen optimally. For example, in the range of $E_b/N_0 = 9$ to 11 dB, the $n = 289$ RCD code with $R_c = 0.83$ achieves the lowest WER(t^*) (from about 10^{-6} to 10^{-13}) amongst the codes considered.

We also observe from Figure IV.8 that, rather interestingly, the behavior of the performance improvement vs. E_b/N_0 is similar to that of WER(t^*). Again, we note that up to about $E_b/N_0 = 9$ dB (the same range where the $n = 49$ code also exhibits the lowest WER(t^*)), the $n = 49$ code manifests the largest improvement, and that the performance improvements for the three longer codes are nearly the same throughout the range of E_b/N_0 plotted. From Figure IV.8, we also observe that for $E_b/N_0 = 9.5$ to 11 dB, the $n = 289$ code achieves, not only the best WER(t^*), but also the most performance improvement amongst the codes considered. Table IV.1(b) is similar to Table IV.1(a), except that the performance improvement of different codes is compared at the same $E_b/N_0 = 10$ dB, instead of the same CSNR. From Table IV.1(b), we see that the code with $n = 289$ achieves the lowest WER(t^*) $= 5.155 \times 10^{-10}$ at $t^* = 0.116$, and also the maximum improvement of 1.98 orders of magnitude, amongst the codes considered.

2.   Fixed $n = 529$, different $d_{\min}$

For the second set of experiments, we plot WER vs. t curves for codes with the same length of 529, but different $d_{\min}$. Figure IV.4 shown previously corresponds to a code of $n = 529$ and $d_{\min} = 6$.

Likewise, Figure IV.9 shows an enlarged version of the WER vs. t plot down to a WER of 10^{-8} for a ($n = 529$, $d_{\min} = 9$) code. A coding gain of ≈ 0.3dB at WER $= 10^{-15}$ is observed if this plot is extended to lower WERs. Similar curves also have been generated for codes with $d_{\min} = 8$, $d_{\min} = 16$, and $d_{\min} = 17$, but are not shown here.

From the plots generated we observe that, for a given CSNR and t, the

Table IV.1

Performance improvement comparisons in employing the optimal BSC/E as opposed to the BSC for codes with (a) $d_{min} = 6$, but different n at CSNR = 10 dB, (b) $d_{min} = 6$, but different n at $E_b/N_0 = 10$ dB, and (c) $n = 529$, but different d_{min} at CSNR = 10 dB.

	n	d_{min}	E_b/N_0 (dB)	CSNR (dB)	t^*	WER(t^*)	WER(0)	Perf. Improvement (orders of magnitude)
(a)	49	6	12.13	10	0.1588	2.1×10^{-16}	1.15×10^{-12}	3.74
	289	6	10.81	10	0.1266	3.9×10^{-13}	2.5×10^{-10}	2.80
	529	6	10.59	10	0.1160	4.81×10^{-12}	1.53×10^{-9}	2.50
	1369	6	10.36	10	0.0995	2.40×10^{-10}	2.66×10^{-8}	2.04
(b)	49	6	10	7.85	0.1446	4.15×10^{-9}	2.53×10^{-7}	1.78
	289	6	10	9.19	0.1160	5.155×10^{-10}	4.90×10^{-8}	1.98
	529	6	10	9.395	0.1068	1.115×10^{-9}	8.49×10^{-8}	1.88
	1369	6	10	9.64	0.093	6.31×10^{-9}	3.0×10^{-7}	1.68
(c)	529	6	–	10	0.1160	4.81×10^{-12}	1.53×10^{-9}	2.50
	529	8	–	10	0.1212	2.87×10^{-16}	8×10^{-13}	3.445
	529	9	–	10	0.1234	1.95×10^{-18}	3.33×10^{-16}	2.23
	529	16	–	10	0.1342	2.37×10^{-34}	8.87×10^{-27}	7.57
	529	17	–	10	0.1350	1.04×10^{-36}	2.04×10^{-30}	6.29

Table IV.2

Coding gain at WER $= 10^{-15}$ due to employing the optimal BSC/E as opposed to the BSC for codes with (a) $d_{\min} = 6$, but different n (b) $n = 529$, but different $d_{\min}$.

	n	$d_{\min}$	WER	Coding Gain (dB)
(a)	49	6	10^{-15}	1.0
	289	6	10^{-15}	0.9
	529	6	10^{-15}	0.8
	1369	6	10^{-15}	0.8
(b)	529	6	10^{-15}	0.8
	529	8	10^{-15}	0.75
	529	9	10^{-15}	0.3
	529	16	10^{-15}	0.5
	529	17	10^{-15}	0.3

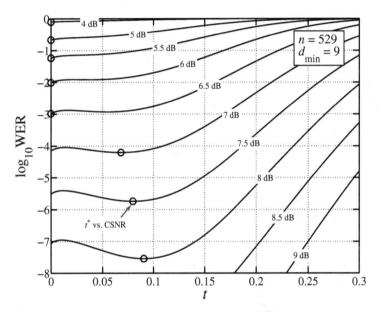

Figure IV.9: For $(n = 529, d_{\min} = 9)$ code: $\log_{10}$WER vs. t vs. CSNR and t^* vs. CSNR.

code with larger $d_{\min}$ achieves a lower WER; this is expected since for the same length a code with larger $d_{\min}$ is stronger in terms of error-erasure correcting capability. These plots, however, indicate a distinct difference in the behavior of WER as a function of t when $d_{\min}$ is odd as compared to when $d_{\min}$ is even. Unlike when $d_{\min}$ is even, from Figure IV.9 we clearly see that at a given CSNR, as t is increased from 0, the WER increases at first. As t is increased further, either the WER reaches a local maximum, then decreases until it achieves a local minimum, as seen at higher CSNR (6, 7, and 8 dB), or the WER simply continues to increase, as seen at lower CSNR (4 and 5 dB). Further, in the event that a local minimum is reached at some $t' \neq 0$, the value of WER at t' may still be higher than the value of WER at $t = 0$. Such behavior is observed at CSNR $= 6$ dB. Consequently, even though a local minimum for the WER $= 0.0122$ is reached at $t' \approx 0.038$, the WER at $t = 0$, which equals 0.0098, is still lower than the value at the local minimum. Thus, $t^* = 0$.

Importantly, the above observations suggest that the existence of a WER local minima on the domain $t \geq 0$ is not guaranteed, in general, to exist when $d_{\min}$ is odd. Thus constrained optimization techniques [45] are required to analytically minimize (IV.2).

Figure IV.10 compares plots of t^* vs. CSNR for codes with $n = 529$ but different $d_{\min}$. The distinction in behavior of codes with odd and even $d_{\min}$ is clearly visible. The value of t^* for codes with odd $d_{\min}$ suddenly drops to 0 as CSNR decreases from a high value. This drop is first due to the local minimum value of $\mathrm{WER}(t) \geq \mathrm{WER}(t = 0)$ for codes with odd $d_{\min}$. As CSNR decreases further, the absence of local $\mathrm{WER}(t)$ extrema, and the monotonically non-decreasing behavior of WER vs. $t \geq 0$, implies that the lowest value of

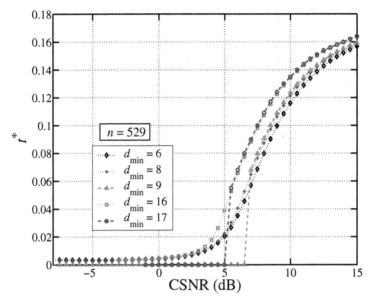

Figure IV.10: t^* vs. CSNR for codes with $n = 529$ but different $d_{\min}$.

WER(t) continues to be achieved at $t = t^* = 0$. In Figure IV.10 at high CSNR ($\gtrsim 7$ dB) the behavior of t^* as a function of CSNR for all different $d_{\min}$ values chosen is similar: at a specific CSNR, the code with larger $d_{\min}$ has a higher t^*. Also, we see that the t^* value for the even-$d_{\min}$ codes decreases smoothly to a non-zero value as CSNR decreases with a very small slope for CSNR $\leq$ 4dB. Additionally, the t^* values for all the even-$d_{\min}$ codes converge as CSNR drops below about 3 dB, which is consistent with the idea of maintaining the average number of erasures at unity (this is a function of only the code length), and the value of $t^* \neq 0$ even at very low CSNR.

Finally Figure IV.11 shows plots of WER (t^*) and $\log_{10}$ WER$(t^*) - \log_{10}$

WER(0) vs. CSNR for $n = 529$ codes with different $d_{\min}$. The plots of WER(t^*) vs. CSNR expectedly show that the larger the $d_{\min}$ of the code (the stronger the code), the lower the WER(t^*). Since R_c is not known for these $d_{\min} = 9, 16$, and 17 codes, we have not plotted t^* and WER(t^*) as a function of E_b/N_0 for the second set of experiments.

In Figure IV.11, the plots of WER(t^*) vs. CSNR, for codes with $n = 529$ and different values of $d_{\min}$, expectedly shows that the larger the $d_{\min}$ of the code, i.e., the stronger the code, the lower the WER(t^*). We have not plotted t^* and WER(t^*) as a function of E_b/N_0 for the second set of experiments, since the code rates are not known for the case when $d_{\min} = 9, 16$, and 17.

From the plots of $\log_{10}$ WER(t^*) $- \log_{10}$ WER(0) vs. CSNR in Figure IV.11, we clearly see that the improvement in performance at any given CSNR for codes with odd $d_{\min}$ is not as much as that with even $d_{\min}$, where the two $d_{\min}$ values are consecutive integers. This observation has already been reported previously in [51]. Additionally, if we consider the different $d_{\min}$ values with the same parity, we see that the code with the larger $d_{\min}$ demonstrates a larger improvement at a given CSNR. For example, at a CSNR of 10 dB, the $d_{\min} = 16$ code has $t^* = 0.1342$ and WER(t^*) $= 2.37 \times 10^{-34}$, while WER(0) $= 8.87 \times 10^{-27}$, giving a performance improvement of about 7.57 orders of magnitude. Corresponding values at the same CSNR, for all the different $d_{\min}$ values considered, are presented in Table IV.1(c), and clearly show that the $d_{\min} = 16$ code has the largest performance improvement of the lot, while the $d_{\min} = 17$ code has the largest performance improvement amongst the codes with odd $d_{\min}$.

Equivalently, the coding gains at a WER of 10^{-15} for the $d_{\min} = 16$ and $d_{\min} = 17$ codes are 0.5dB and 0.3dB, respectively. Table IV.2(b) shows the coding gain in dB for the codes with $n = 529$ but different $d_{\min}$. From Table IV.2, it is see that:

1. coding gain decreases with n for fixed $d_{\min}$.

2. coding gain decreases with $d_{\min}$ for fixed n and when the parity of $d_{\min}$ (even or odd) is kept fixed.

E. Extended-BFA Decoding for RCD Codes on the BSC/E

In the previous section we focused our attention on the performance of codes on the BSC/E based on the extended-BDD decoder. For the class of regular LDPC codes (including RCD codes), we know that the BFA discussed in Chapter III performs much better than the BDD. Further, the extension algorithm for the BSC/E presented in Section IV.B can be used in conjunction with any constituent HDD. Consequently, for RCD codes on the BSC/E, one would certainly expect that the extended-BFA (e-BFA) would perform better than the extended-BDD.

In this section, we discuss extended-BFA decoding of RCD codes on the BSC/E.

One important difference in the analysis of the e-BFA algorithm as compared to the analysis of the BFA is that one cannot use the performance obtained by transmission of the all-zeros codeword as a representative of code performance in general. This can be easily understood from the following argument. Before doing so, we point out that for the analysis in this section, we consider only

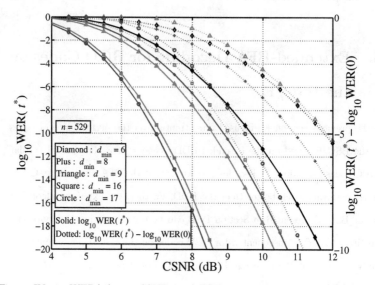

Figure IV.11: WER(t^*) vs. CSNR, and WER improvement between BSC/E $(t = t^*)$ and BSC $(t = 0)$ vs. CSNR, for codes with $n = 529$ but different $d_{\min}$.

logical values in the e-BFA decoder (i.e., a logical-0 and a logical-1). Thus, in the first CD of the e-BFA, an erasure is replaced by a logical-0 (instead of replacing the erasure with a '1'), while in the second CD of the e-BFA, the erasure is replaced by a logical-1 (instead of replacing the erasure with a '-1'). Henceforth, the first CD also may be referred to as the 0-path and the second CD as the 1-path (for obvious reasons).

Assume that a randomly-chosen codeword is transmitted on the BSC/E and it gets corrupted by an e-error f-erasure pattern. For convenience in explanation, let us assume that $e = 0$. When the error-erasure pattern is input to the first CD, all the erasures are set to 0. If the transmitted codeword was all-zeros, the input to the first CD will be the all-zeros codeword and, clearly, the first CD will decode successfully to the all-zeros codeword irrespective of the value of f. On the other hand, for the randomly-chosen transmitted codeword, on average $f/2$ of the erased positions are likely to have a 1 in the transmitted codeword, while on average the other $f/2$ positions are likely to have a value of 0 in the transmitted codeword. Thus, for a randomly-chosen transmitted codeword, setting all f erasure positions to 0 will result in an error-pattern of average weight $f/2$ being added to the transmitted codeword before being input to the first CD (and also, similarly, to the second CD). This means that there is no guarantee that at least one CD will decode to the randomly-chosen transmitted codeword when $f/2 > \tau$, where τ is the error-correcting capability of the decoder.[2]

---

[2]Of course, if $f/2 \leq \tau$, then at least one CD will decode to the transmitted codeword irrespective of which codeword is transmitted. We use the variable τ to avoid confusion with the BSC/E threshold t.

Also note that when the transmitted codeword is the all-zeros codeword the input to the second CD is an error pattern of weight f. Thus, the second CD is not very likely to decode to the all-zeros codeword. In any event, the output of the second CD will be ignored if it gives a decoder failure, in which case the output of the e-BFA will be the output of the first CD, i.e., the all-zeros codeword. In fact, the only case when the output of the e-BFA will be different from the output of the first CD will be when the second CD decodes to a valid non-zero codeword, in which case the e-BFA must signal a decoder failure. Such events, however, are likely to be relatively infrequent.

It is obvious that the procedure of setting all the erasures to zeros and ones, respectively biases the e-BFA towards choosing the output of the first CD (the all-zeros codeword) over choosing the output of the second CD when the transmitted codeword is all-zeros. For randomly-chosen codewords, on the other hand, such a bias does not exist on an average. When the transmitted codeword is the all-zeros codeword, the output of the e-BFA is very likely the output of the first CD, which is the all-zeros codeword. This, of course, corresponds to successful decoding. Consequently, we must conclude that the e-BFA favors correct decoding more strongly when the all-zeros codeword is transmitted as compared to when a randomly-chosen codeword is transmitted. This, in turn, implies that the e-BFA will achieve lower WER if one considers only the transmission of the all-zeros codeword than if one considered the transmission of randomly-chosen codewords.

The inability to assume all-zeros codeword transmission complicates the analysis of decoding of error-erasure patterns when $2e+f \geq d_{\min} = 6$. For $2e+f < 6$, however, one can show that the e-BFA can correct all possible such e-error and

f-erasure patterns, even if one assumes the transmission of randomly-chosen codewords.

1. Decoding of e-error f-erasure patterns that satisfy $2e + f \leq 5$

First, we prove a useful lemma that establishes the Hamming weight between an error-erasure pattern and an arbitrary codeword under consideration. Next, we show that for RCD codes the e-BFA can correct all error-erasure patterns that satisfy $2e + f \leq d_{\min} - 1$.

Lemma IV.1: *For an e-error and f-erasure pattern and an arbitrary code with a codeword at weight $d \geq e + f$, the Hamming distance h between the codeword and the error-erasure pattern satisfies $h \geq d - e$, with equality achieved if and only if the errors and erasures occur at positions that are contained within the set of 1-positions in the codeword of weight d.*

Proof: Since the erasure locations always contribute to the Hamming distance, $h \geq f$. The codeword has d 1-positions. The error-erasure pattern has e 1-positions and f erasure-positions. If the 1-positions and erasure-positions for the error-erasure pattern are chosen from the set of 1-positions for the codeword (this is possible since $d \geq e + f$), then there will be exactly $d - e$ positions where the symbol in the codeword will differ from the symbol in the error-erasure pattern. Now let the error-erasure pattern be such that e_{out} 1-positions and f_{out} erasure-positions are located outside the set of 1-positions of the codeword. Then

$$h = d - (e - e_{\text{out}}) + e_{\text{out}} + f_{\text{out}} = d - e + 2e_{\text{out}} + f_{\text{out}}. \qquad \text{(IV.9)}$$

Clearly $h = d - e$ if and only if $e_{\text{out}} = f_{\text{out}} = 0$, i.e., when the errors and erasures occur at positions that are contained within the set of 1-positions in the codeword of weight d. $\square$

Theorem IV.1: *For an RCD code of length $n = \eta^2$ (with column weight $\gamma = 3$), all e-error and f-erasure patterns satisfying $2e + f \leq d_{\min} - 1 = 5$ can be successfully corrected by the e-BFA.*

Proof: Assume the transmission of an arbitrary codeword. When the error-erasure pattern has only errors, the inputs to the two CDs and, consequently, the outputs of the two CDs will be identical. The only non-negative values of e that satisfy $2e + f \leq 5$ are 0, 1, and 2. From Theorem III.2 we know that the CD (the BFA, in this case) is capable of successfully decoding all e error-patterns for $e \leq 2$.

Now let us consider a situation when the error-erasure pattern has a nonzero number of erasures and satisfies $2e + f \leq 5$. Let $f = 1$. If the transmitted codeword has a 1 at the erased position, the input to the second CD (1-path, where erasures are set to logical-1s) will be an error pattern of weight $e \leq 2$, which will be successfully decoded to the transmitted codeword. On the other hand, the input to the first CD (0-path, where erasures are set to logical-0s) will be an error pattern of weight $e + f \leq 3$, which may or may not decode to the transmitted codeword. If the first CD decodes to a valid codeword other than what was transmitted, the Hamming distance between the output of the first CD and the received word will be at least $d_{\min} - e \geq 4$, since this codeword is at a distance of at least $d_{\min} = 6$ away from the transmitted codeword (also see Lemma IV.1). On the other hand, the Hamming weight between the received word and the output of the second CD equals $e + f \leq 3$. Thus, recalling the decision mechanism of the extension algorithm, we see that no matter what the output of the first CD is, the output of the second CD – the transmitted codeword – will always be as the e-BFA output. A similar argument can be applied to the case where the transmitted codeword has a 0 at the erased position, to show the the output of the e-BFA will again be the transmitted codeword, although this time the transmitted codeword is guaranteed to be the output of the first CD.

In general, let f_0 equal the number of positions in the transmitted codeword that have zeros in the erased locations, while let f_1 equal the number of positions in the transmitted codeword that have ones in the erased locations. Clearly, $f_0 + f_1 = f$ and $0 \leq f_0, f_1 \leq f$. The input to the first CD will be an error pattern of weight $e + f_1$, while the input to the second CD will be an error pattern of weight $e + f_0$. Since $2e + f \leq 5$, we get $e + f_0 + e + f_1 \leq 5$. A little thought reveals that at least one of the two following conditions is always satisfied: (a) $e + f_0 \leq 2$, (b) $e + f_1 \leq 2$. If condition (a) is satisfied, the second CD is guaranteed to decode to the transmitted word, while if condition (b) is satisfied, the first CD is guaranteed to decode to the transmitted word.

Let us now assume that condition (a) is satisfied and (b) is not. In this case, if the first CD also decodes to the transmitted codeword or gives a decoder failure, we are assured of choosing the output of the second CD, i.e., the transmitted codeword as the output of the e-BFA. The situation that concerns us most is when the first CD decodes to a codeword other than what was transmitted. When this happens, the Hamming distance between the received word and the output of the first CD is at least $6 - e$ from Lemma IV.1, whereas the Hamming distance between the received word and the output of the second CD is $e + f$. Of course, since $2e + f < 6$, we have $e + f < 6 - e$. Thus, the output of the second CD – the transmitted codeword – will always be chosen if condition (a) is satisfied alone. The discussion for the case when condition (b) is satisfied, but condition (a) is not, proceeds similarly. Finally, if both condition (a) and (b) are satisfied, both the CDs will decode to the transmitted codeword. Thus, decoding to the transmitted codeword is assured in all cases if $2e + f \leq 5$. $\square$

2. Decoding of e-error f-erasure patterns that satisfy $2e + f = 6$

Next, we focus on the decoding of error-erasure patterns that satisfy $2e + f = 6$ under the assumption that the all-zeros codeword is transmitted. The analysis of error-erasure patterns that satisfy $2e + f = 6$ for arbitrary codeword transmission turns out to be more complicated. As we argued previously, the e-BFA WER performance under all-zero transmission will be better than the average WER performance for a randomly-chosen codeword. Thanks to this, any lower bound on the WER of the e-BFA under the all-zeros codeword assumption will necessarily serve as a lower bound on the average WER of the e-BFA, although this lower bound may not be tight. The analysis of $2e + f = 6$ error-erasure patterns under the all-zeros codeword assumption enables us to compute lower bounds for performance under this assumption. This lower bound, in turn, also serves as a lower bound on average e-BFA WER.

The set of all error-erasure patterns satisfying $2e + f = 6$ can be partitioned into four different cases, which are: (a) $e = 3, f = 0$, (b) $e = 2, f = 2$, (c) $e = 1, f = 4$, and (d) $e = 0, f = 6$. In case (a) the input to both the CDs is the same 3-error pattern. The BFA decoding of 3-error patterns has been discussed at length in Chapter III and all the results obtained there can be directly used to analyze this case. For the other three cases, since $f > 0$, the inputs to the two CDs will be different. Consequently, the outputs of the two CDs may also differ. In the following theorem, under the assumption of the all-zeros codeword transmission, we show that for the three remaining cases, the output of the first CD is always the all-zeros codeword. Given this, the only possibility of the e-BFA not decoding to the all-zeros codeword occurs when the second CD decodes to a codeword other than the all-zeros codeword and $h_2 \leq h_1$; here h_1 and h_2 are the Hamming distance between the e-BFA input error-erasure pattern and the output of the first CD, and the Hamming distance between the e-BFA input error-erasure pattern and the output of the second CD, respectively.

Theorem IV.2: *For an RCD code of length* $n = \eta^2$, η *odd, e-error and f-erasure patterns satisfying* $2e + f = d_{\min} = 6, f > 0$, *under the assumption of all-zeros codeword transmission and decoding via the e-BFA, exactly* $A_6 \binom{6}{f}\binom{6-f}{e}$ *patterns lead to decoder failure, while all other patterns get successfully decoded to the all-zeros codeword. Here* $A_6 = \eta \binom{\eta}{3}$ *is the number of codewords at* $d_{\min} = 6$.

Proof: The input to the first CD will be an error pattern with weight equal to e. Since $f > 0$, then $e \leq 2$, and applying Theorem III.2 we see that in all cases the first CD will decode to the all-zeros codeword and the Hamming distance between the e-BFA input error-erasure pattern and the output of the first CD equals $h_1 = e + f$. Using arguments similar to those presented in Theorem IV.1, we see that the input to the second CD will be an error-pattern with weight $e + f > e$. In the event that the second CD decodes to a codeword at weight $d \neq 0$, the Hamming distance between the e-BFA input error-erasure pattern and the output of the second CD must satisfy $\min (h_2) = d - e$ (see Lemma IV.1). The e-BFA will not decode to the all-zeros codeword only if the second CD decodes to a codeword other than the all-zeros codeword and $h_2 \leq h_1$.

One situation when $h_1 < \min(h_2)$, i.e., $2e + f = 6 < d$, is satisfied is when $d > 6$. In other words, if the second CD decodes to a codeword of weight > 6, then $h_1 < \min(h_2)$ and the output of the e-BFA will be the all-zeros codeword. We also note that $h_2 < h_1 \Rightarrow d - e < e + f \Rightarrow d < 2e + f = 6$, which is not possible. Thus, we can further say that for error-erasure patterns with $2e + f = d_{\min} = 6, f > 0$, the e-BFA will not decode to the all-zeros codeword only if the second CD decodes to a codeword of weight 6 and $h_2 = h_1 = e + f$.

From Lemma IV.1 we see that $h_2 = h_1 = e + f$ is possible only when the errors and erasures occur at positions that are contained within the set of 1-positions in the codeword of weight 6 to which the error-erasure pattern decodes. Using elementary combinatorics, we see that there are exactly $\binom{6}{f} \binom{6-f}{e}$ such e-error and f-erasure patterns for each codeword[3]. Further, since $f > 0$, a $2e + f = 6$ error-erasure pattern that is contained within a codeword $\mathbf{c_1}$ at weight 6 will always be different from a similar pattern that is contained within another codeword $\mathbf{c_2}$ at weight 6. This follows since: (a) If $f > 0$ and $2e + f = 6$ then $e + f \geq 4$, i.e., there are at least four positions with either errors or erasures, and (b) $\mathbf{c_1}$ and $\mathbf{c_2}$ can have at most three 1-positions in common (if not, the Hamming distance between the two codewords $\mathbf{c_1}$ and $\mathbf{c_2}$ will be ≤ 4, which is not possible, since the code minimum-distance is 6). Thus, since there are A_6 codewords at weight 6, the total number of distinct error-erasures patterns with $f > 0$, $2e + f = 6$ that are contained within the set of 1-positions in a codeword of weight 6 is exactly $A_6 \binom{6}{f} \binom{6-f}{e}$. $\square$

---

[3]Consider an e-error f-erasure pattern with errors and erasures occurring at positions that are contained within the set of 1-positions in a codeword of weight 6. The erasures may be

From Theorem IV.2, for an RCD code of length η^2 under e-BFA decoding, the exact number of e-error and f-erasure patterns satisfying $2e+f = d_{\min} = 6, f > 0$ that decode successfully to the all-zeros codeword can be computed as

$$z_{e,f} = \binom{\eta^2}{f}\binom{\eta^2 - f}{e} - A_6\binom{6}{f}\binom{6 - f}{e}. \qquad \text{(IV.10)}$$

Also, for all the other $A_6\binom{6}{f}\binom{6-f}{e}$ patterns, the e-BFA declares a decoder failure. Here, it is worthwhile to point out that the result of Theorem IV.2 also has been verified by exhaustive generation, enumeration, and e-BFA decoding of error-erasure patterns satisfying $2e + f = d_{\min} = 6, f > 0$, for RCD codes with $\eta = 5, 7, 9$.

Equation (IV.10) along with the expression for z_3 (the number of 3-error patterns that decode successfully to the all-zeros codeword for RCD codes under BFA decoding) from equation (III.27) can be used to construct lower and upper bounds on the decoding performance of RCD codes under the e-BFA with the assumption of all-zeros codeword transmission. In particular, the WER of an RCD code of length $n = \eta^2$ under e-BFA decoding with all-zeros codeword transmission can be upper-bounded as

$$
\begin{aligned}
\text{WER}_{\text{e-BFA},0} \quad \leq \quad & 1 - \sum_{\substack{e,f\geq 0; \\ (2e+f)\leq 5}} \binom{n}{f}\binom{n - f}{e}\alpha^f \beta^e (1 - \alpha - \beta)^{n-e-f} \\
& - \sum_{\substack{e,f\geq 0; \\ (2e+f)=6}} \gamma_{e,f}\binom{n}{f}\binom{n - f}{e}\alpha^f \beta^e (1 - \alpha - \beta)^{n-e-f},
\end{aligned}
$$
$$\text{(IV.11)}$$

---

located at f of the six 1-positions in $\binom{6}{f}$ ways; the errors may be located at e of the remaining $6 - f$ available positions in $\binom{6-f}{e}$ ways.

where

$$\gamma_{e,f} = \frac{z_{e,f}}{\binom{n}{f}\binom{n-f}{e}} = 1 - \frac{A_6\binom{6}{f}\binom{6-f}{e}}{\binom{n}{f}\binom{n-f}{e}}. \qquad \text{(IV.12)}$$

Note also that $z_{3,0} = z_3$.

Similarly, the WER may also be lower-bounded as

$$\text{WER}_{e-\text{BFA},0} \geq \sum_{\substack{e,f \geq 0; \\ (2e+f)=6}} (1 - \gamma_{e,f}) \binom{n}{f}\binom{n-f}{e} \alpha^f \beta^e (1 - \alpha - \beta)^{n-e-f}, \qquad \text{(IV.13)}$$

Since the average WER performance of the e-BFA is worse than under the all-zeros codeword assumption, equation (IV.11) is not of much use in bounding the average WER performance. Equation (IV.13), however, also serves as a lower bound on the average WER performance of the e-BFA.

3. Performance curves for some RCD codes

Figure IV.12 shows the various performance curves for the $\eta = 23$ RCD code on the BSC/E, where the BSC/E was generated based on symmetric decision thresholds at $\pm t$, with $t = 0.116$. This value of t was chosen since it gave the lowest WER based on the decoder's ability to correct all error-erasure patterns satisfying $2e + f \leq 5$ at $E_b/N_0 = 10.6$dB (see Table IV.1).

The dotted curve (blue) shows the WER of this code based on BDD for the BSC. The solid curves (red) indicate the upper and lower bounds on the WER based on BFA for the BSC. These bounds are computed via equations (III.37) and (III.39). The circles (red) show the WER from standard MC simulations for the $\eta = 23$ RCD code based on BFA decoding on the BSC. These curves are discussed in detail in Chapter III and are provided here for reference.

The dashed curve (blue) shows the upper bound on the WER of the e-BDD algorithm based on successful decoding of all e-error f-erasure patterns that satisfy $2e + f \leq 5$. In plotting this curve, it is assumed that no error-erasure pattern satisfying $2e + f \geq 6$ is decoded successfully. Since the e-BFA is also capable of correcting all e-error f-erasure patterns that satisfy $2e + f \leq 5$ (see Theorem IV.1), this curve also serves as an upper bound (albeit loose) on the WER of the e-BFA. The two dash-dot curves (black) indicate upper and lower bounds on the WER of the e-BFA that are computed via equations (IV.11) and (IV.13), respectively, and based on the all-zeros codeword transmission. The $\times$ symbols represent the WER from standard MC for the $\eta = 23$ RCD code using e-BFA decoding on the BSC/E and based on the transmission of the all-zeros codeword, while the squares with the dotted line represent WER from standard MC simulations for the $\eta = 23$ RCD code using e-BFA and based on the transmission of randomly-generated codewords. This curve thus represents the average WER performance of the $\eta = 23$ RCD code on the BSC/E under e-BFA decoding. All standard MC estimates for the BSC/E correspond to the collection of 100 word-error events.

The figure substantiates our previous argument that the performance of the RCD code with e-BFA decoding on the BSC/E is superior (a gain of about 0.5 dB at WER $= 10^{-4}$) under the assumption of all-zeros codeword transmission than under the assumption of randomly-chosen codeword transmission. The gain under the assumption of all-zeros codeword transmission, however, is seen to decrease with E_b/N_0. Also, the lower bound on e-BFA performance obtained assuming all-zeros codeword transmission still serves as a lower bound on the average e-BFA performance of the RCD code.

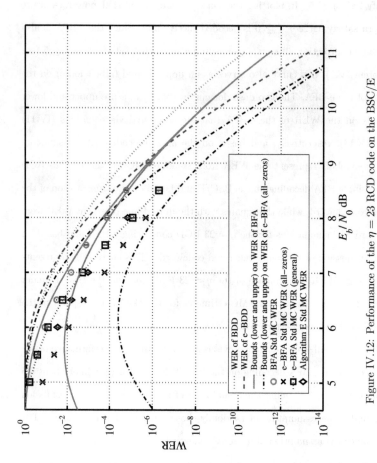

Figure IV.12: Performance of the $\eta = 23$ RCD code on the BSC/E

164

From Figure IV.12, it also is evident that substantial performance gains are possible by using a BSC/E model with an appropriately chosen threshold instead of a BSC model. For example, at WER $= 10^{-6}$, a gain of about 0.6dB is (horizontal separation between the two merged solid curves and the dotted curve with squares) is observed in communicating via a BSC/E with $t = 0.116$ instead of a BSC. This gain also appears to increase with decreasing WER. At WER $= 10^{-10}$, a gain of about 0.8dB is expected (horizontal separation between the two merged solid curves and extrapolation of the dotted curve with squares). Note that one cannot use the upper dash-dot curve as a measure of average e-BFA WER, due to reasons discussed previously.

Finally, we also compare the performance of e-BFA decoding with decoding based on algorithm E given in [65], which is a message-passing decoding algorithm for the BSC/E. For algorithm E, code performance evaluation under the all-zeros codeword transmission assumption is indeed representative of the average code performance, since all message-passing decoders satisfy the symmetry conditions [65]. The performance based on decoding via algorithm E is shown by the diamonds in Figure IV.12. We observe that at low E_b/N_0 the average WER performance of algorithm E is better than the average e-BFA performance. At about 8 dB, however, the average e-BFA performance is better than the average performance of algorithm E. We note that the chosen threshold of $\pm t = 0.116$ may not be the best threshold for both the e-BFA and algorithm E ($\pm t = 0.116$ is optimal for the e-BDD at 10.6 dB) and, hence, both decoders could perform better. Also, we chose a weight sequence of $\{2, 1, 1, \ldots, 1\}$ (see [65] for a discussion of this weight sequence) for the 20 iterations of algorithm E, which may not be the optimal weight sequence for the $\eta = 23$ RCD code.

F. Conclusions

1. Summary and conclusions

In this chapter, we studied the word-error-rate (WER) performance of forward error correction (FEC) codes, with different lengths n and minimum distances $d_{\min}$, on the binary symmetric channel with erasures (BSC/E) and decoding via an extension of BSC hard-decision decoders; the performance of our RCD codes was of particular interest in this study and motivated the specific choices of code parameters that were studied. The BSC/E is assumed to be generated by a binary phase-shift-keying (BPSK) additive white Gaussian noise (AWGN) channel with variance σ^2 and a pair of symmetric decision thresholds at $\pm t$. Plots were presented for the WER as a function of the BSC/E decision threshold t for different channel signal-to-noise ratios (CSNR $\triangleq 10 \log_{10} \left(\frac{1}{2\sigma^2}\right)$) for the extension of the bounded-distance decoder (BDD) to the BSC/E.

We addressed the minimization of WER with respect to t, where all other variables were kept fixed. For given code parameters, the WER-minimizing thresholds t^* and the minimum WER (WER(t^*)) were evaluated at different CSNR and, where possible, at different E_b/N_0 and plotted. While the initial theory was developed under the general assumption of extending any hard-decision decoder to the BSC/E, plots and results were presented for only the extended BDD (e-BDD). Finally we studied the performance of RCD codes on the BSC/E with decoding via the extension to the BFA (e-BFA). Some bounds on e-BFA performance for RCD codes were derived, and plots of e-BFA performance were presented for the $\eta = 23$ RCD code. The WER performance of the e-BFA also was compared with that of a message-passing decoder for the BSC/E.

We observe that substantial improvements in WER performance of RCD codes, as well as for FEC codes in general, can be obtained, especially at higher CSNR values, by employing two thresholds instead of one threshold, with the thresholds chosen to minimize WER. At a specific CSNR, these improvements increase with increasing d_{min} for a fixed length, but decrease with increasing length for a fixed d_{min}, suggesting that the stronger codes benefit more from softer decoding. On the other hand, at a specific WER, the coding gain in dB decreases with n for fixed d_{min}, and decreases with d_{min} for fixed n when the parity of d_{min} (even or odd) is kept fixed. From this viewpoint, thus, the stronger codes benefit lesser from decoding using erasures. Nevertheless, the coding gains obtained for the codes considered in this chapter are significant to merit the use of a BSC/E instead of a BSC.

A distinct difference is observed between codes with odd and even d_{min} in the behavior of WER as a function of t, and of t^* as a function of CSNR. Codes with even d_{min} such as RCD codes have a larger improvement in performance in going to a BSC/E from a BSC as opposed to codes with odd d_{min}. This is primarily due to the fact that the ability of even d_{min} codes to correct an extra erasure which was not utilized on the BSC becomes available on the BSC/E.

For a collection of codes on the BSC/E with the same d_{min} but different lengths, under the assumption that each code operates with the optimal threshold, each code is optimal for some sub-interval of E_b/N_0. Further, in such a case, for every E_b/N_0 value one can find a code that provides the best WER performance among codes in that collection. For codes with the same length but different even d_{min}, the WER-minimizing thresholds are seen to converge to a non-zero value at very low CSNR. On the other hand, for codes with odd d_{min}, the best performance at very low CSNR is achieved by setting $t = t^* = 0$ (i.e., by employing the BSC).

Based on the discussion presented, we conclude that, in general, the existence of a local minima for the WER of the e-BDD as a function of $t \geq 0$ is not guaranteed when $d_{\min}$ is odd. Furthermore, even if a local minimum exists, it need not be a global minimum. Thus, analytical minimization of the WER of the e-BDD as a function of $t \geq 0$ will require constrained optimization techniques. In fact, in all examples of codes with even $d_{\min}$ considered (which includes the RCD codes), we have always observed the existence of $t^* \neq 0$. Conclusive proof of this fact for all even $d_{\min}$, however, remains to be demonstrated.

We also conclude that for the $\eta = 23$ RCD code, a coding gain of about 0.8dB may be obtained at a WER of 10^{-10} in communicating via a BSC/E with $t = 0.116$ and e-BFA decoding, instead of communicating via the BSC with BFA decoding. Further, this coding gain increases as WER decreases. Also, performance of the e-BFA under the assumption of transmitting the all-zeros codeword is not an accurate measure of true e-BFA performance (i.e., with transmission of randomly-chosen codewords). Further, e-BFA decoding may provide better performance than that provided by a message passing decoder (algorithm E of [65]) for the BSC/E.

2. Future research

1. In general, it is not sufficient to analyze the decoding performance of codes based on the e-BFA presented in this chapter by restricting ourselves to the transmission of the all-zeros codeword. Hence, a combinatorial analysis of e-error f-erasure patterns for RCD codes satisfying $2e + f = 6$ that assumes

the transmission of an arbitrary codeword would be very much desirable.

2. Modifications to the extension algorithm may be investigated with the objective of improving decoder performance. One particular suggestion involves the use of more constituent decoders (CDs), where the input to each CD is an error-pattern obtained by a unique replacement of erasures with ones and zeros.

3. A comparison between the e-BFA performance and algorithm-E performance of RCD codes at a specific high value of E_b/N_0, with each decoder operating at its optimal BSC/E decisioning threshold at that E_b/N_0, is very much in order.

4. A thorough comparison of the e-BFA algorithm and algorithm E of [65] in terms of computational complexity is certainly desirable for assessing RCD code utility. We note that the BFA (the CD for the e-BFA) requires only a single computation to be performed at each parity-check node of the bipartite graph (an XOR operation). Algorithm E, on the other hand, requires ρ different computations (ρ is the row weight of the regular LDPC code parity-check matrix) to be performed at each parity-check node (one for each outgoing message that must be transmitted). Similarly, γ different computations are required at each variable node for algorithm E. The BFA, on the other hand, requires only a single computation at each variable node, followed by a search over all variable nodes to find the maximum. Also, the complexity of the e-BFA can be assumed to be roughly twice the complexity of the BFA (if one ignores the complexity of the decisioning that needs to be performed once both CD outputs are available).

5. It would be very useful to obtain an ASIC and/or FPGA implementation for the e-BFA decoder for certain RCD codes (such as the $\eta = 23$ RCD code) in order to evaluate RCD code performance on physical systems, e.g., optical fiber communications systems.

6. Our motivation to investigate the BSC/E arose from the desire to improve the performance of communications systems such as the optical fiber communications (OFC) system using RCD codes, without an excessive increase in computational complexity. Further, it has also been shown [62,65] that for LDPC codes, a major portion of the coding gain that can be achieved in going from a BSC with hard-decision iterative decoding to a BPSK-AWGN channel model with soft-decision iterative decoding occurs in simply going to the BSC/E. The OFC channel, however, is inherently asymmetric and frequently modeled with chi-squared *pdfs* leading to a binary asymmetric channel (BAC). With this in mind, we desire to extend our analysis to tackle the BAC and variations of the binary asymmetric channel with asymmetric erasures (BAC/AE) for RCD codes specifically, and LDPC codes in general.

Chapter V.

EVALUATION OF VERY LOW BER OF RCD CODES USING DUAL ADAPTIVE IMPORTANCE SAMPLING

In this chapter we evaluate the performance of some RCD codes down to low word-error rates (WERs) of the order of 10^{-15} on a binary phase-shift-keying (BPSK), additive white Gaussian noise (AWGN) channel with soft-decision iterative decoding based on the sum-product algorithm (SPA) [41]. Conventional simulation techniques such as standard Monte Carlo [36] are found to be incapable of estimating such low WERs in reasonable time. Thus, alternative techniques must be employed to achieve our objective. The technique of our choice is the dual-adaptive importance sampling (DAIS) technique originally conceived and developed by Holzlöhner, Mahadevan, and others [33] for the coded communications problem.

From a practical stand point, the desire to evaluate RCD codes to low WER and bit-error rate (BER) values is primarily motivated by the fact that these codes are possible candidates for deployment in optical fiber communications (OFC) systems, which are currently required to operate at design BERs in the range of 10^{-12} to 10^{-15}. From a theoretical stand point, the evaluation of RCD code performance to low WER and BER is desired as it enables us to compare code performance with existing bounds such as the union bound [15], which is known to be tight at higher signal-to-noise ratios (SNRs). For codes in general, for which bounds may not be known, low BER/WER evaluations of performance can help determine the presence of error floors that may affect code performance at lower error rates (or higher SNRs).

The outline of this chapter is as follows. We first provide some introduction and background to the performance evaluation of FEC codes, especially to low values of probability of error. We also provide some background on the various *importance sampling* (IS) techniques used to evaluate codes to low WER and BER, and introduce the *dual-adaptive importance sampling* (DAIS) technique. We briefly discuss reasons for choosing DAIS as a vehicle to evaluate RCD codes. Then we present the key underpinnings of the DAIS technique in a general framework. The reader is referred to [33] for a more detailed description of the application of the algorithm to evaluate code performance on a specific coded communication system. Next, we use a regular $n = 96$, $k = 50$ LDPC as an example to demonstrate the use of DAIS. Then we present and discuss the performance curves for three different RCD codes, which is the most important part of this chapter. Finally, we end with the summary, conclusions, and future research.

A. Background and Introduction

The performance of forward error correction (FEC) codes in digital communication systems is almost exclusively studied and compared on the basis of performance curves – plots of bit-error rate (BER) and/or word-error rate (WER) versus the information signal power to noise power ratio (SNR). The exact analytical evaluation of FEC code performance curves under a specific channel and decoding scheme is extremely difficult to perform and computationally intractable, even for moderately long codes, say with more than 30 to 40 information bits per codeword.

Approximate performance of FEC codes under certain decoding schemes may be obtained analytically in the form of lower and upper bounds, provided knowledge of the weight enumerator function (WEF) of the code is available; partial knowledge of the WEF in the form of the first non-trivial term may also suffice [23,75]. In general, these bounds serve as very good approximations to the actual code performance at high SNR values. The quality of the approximations, however, is not very good at relatively moderate-to-low values of SNR. To make matters worse, unless the code belongs to one of the few classes of codes whose WEF is already known, such as the RS and BCH codes that still abound in practical FEC schemes, the computation of the code WEF or even the first non-trivial term is itself intractable for codes of practical size [23].

Thus, in general, for a large set of codes with parameters in the realm of current practical applicability (100s and 1000s of bits long), it is impossible for FEC code researchers and developers to analytically obtain performance curves, either exact or in the form of bounds. Consequently, for some time now, the preferred methodology employed by communications engineers and researchers to study code performance, has and continues to be based on performing Monte Carlo computer simulations or physical system experiments.

Monte Carlo simulations and physical system experiments, which we will collectively call standard Monte Carlo techniques, have been quite effective in obtaining performance curves down to BER values as low as about 10^{-7}. Attempting to extend the performance curves to lower values of BER, however, continues to be a challenge even with current computing capabilities because of the inordinate amounts of time required – on the order of months or years – to obtain meaningful results at such low BER values [18,46].

Technological advancements in the very recent past have drastically altered the scenario in the communications industry. The industry has witnessed an explosion in communications speeds and capacity of storage media, in addition to improvements in the overall quality of communications; this has led service providers to impose more stringent requirements on the error rate of communication systems, e.g., optical fiber communication (OFC) systems of today operate at design BERs in the range of 10^{-12} to 10^{-15}. Unfortunately, standard Monte Carlo techniques prove to be woefully inadequate for FEC codes and decoders at such low BER values. In an attempt to extend the range of standard Monte Carlo techniques and provide some assistance in the evaluation of higher performing systems, researchers have called upon the services of supercomputers or system experiments. Both of these approaches are characterized by unduly long performance time requirements, and even then, the best reliable BERs that have been achieved are in the order of 10^{-10} or 10^{-11} [18,46,71] and are acquired over time spans measured in months. This is very expensive to industry and researchers in terms of person-hours and computer time.

One of the most promising research areas in the context of evaluating FEC code performance to low WER and BER values, much below that capable via standard Monte Carlo techniques, has been the study of *importance sampling* (IS) techniques [2,36]. Importance sampling is based on increasing the frequency of occurrence of very low probability (but important) error events defined by the channel noise probability density function (*pdf*) by using a related *modified* or *biased* *pdf* to increase the production of such events. The choice of a suitable biased *pdf* is of critical importance to the success of IS techniques. An optimal biased *pdf* for IS does exist in theory. This optimal biased *pdf*, however, is heretofore not realizable since its construction requires knowledge of the very probability that we desire to estimate. Consequently, biased *pdfs* are empirically chosen, usually based on some knowledge of the regions in the sample space where the event of interest E (the event whose probability $P(E)$ we desire to estimate) is likely to occur. In the context of FEC code and decoder performance evaluation, this boils down to choosing a biased *pdf* using some knowledge of which noise realizations are most likely to generate errors at the high SNR values where the low BERs must be estimated.

Most recently proposed techniques for evaluating FEC codes via importance sampling [22, 66, 77, 79] rely on using some code-property to obtain a rough estimate of the region in noise space that is a major contributor to the BER; e.g., Sadowsky [66], as well as Ferrari and Bellini [22], exploit the knowledge of the minimum-weight codewords, to determine a suitable biased *pdf*. Unfortunately, as we have discussed, this knowledge is practically unobtainable for codes in general.

Xia and Ryan [77] use the regular structure of the bipartite graphs of certain codes to determine the biased *pdf*. For every codeword that is received, the BER evaluation using the IS technique proposed by them is performed for only a single bit-position in the codeword. Standard Monte Carlo techniques, on the other hand, evaluate the BER for all n bit-positions for every received codeword of length n. Hence for the transmission and decoding of the same number of codewords (which can be assumed to take the same time for both IS as well as standard Monte Carlo techniques), the standard Monte Carlo technique evaluates n times as many bits as their IS technique. This is what Xia and Ryan refer to as the 'divide-by-n' problem. According to the authors' analysis, this divide-by-n problem is likely to affect the efficiency of their proposed IS technique relative to the standard Monte Carlo techniques for codes with large n. In addition, the proposed IS technique is predicated on the regular structure in the bipartite graph of codes, which justifies the analysis of a single bit-position in the code to estimate the BER of the entire code. A regular bipartite graph structure is obtained from regular low-density parity-check (LDPC) codes, but is usually not consistent with irregular LDPC codes or codes in general.

The task of determining an appropriate biased *pdf* is FEC code specific, and, as we can see, while it may be possible to choose biased *pdfs* for certain specific codes or classes of codes, information on how to appropriately bias is not available for codes in general.

The *multicanonical Monte Carlo* (MMC) technique, an adaptive technique, attempts to mitigate the problem of choosing a suitable biased *pdf* by iteratively converging to it. Like standard importance sampling, MMC increases the number of events in the tail of the original *pdf* by sampling from a biased *pdf* [34]. The advantage of the MMC method is that it automatically determines the biased *pdf* (that approximates the optimal biased *pdf*) by using an iterative procedure that requires relatively little a priori knowledge of how to bias. The iterative procedure uses a *control variable* to update the biased *pdf* for the next iteration and a Metropolis random walk [57] to sample from the biased *pdf*.

While the MMC algorithm may be satisfactory to estimate the $P(E)$ in many cases, it has been found from simulations to be inadequate to evaluate the error correction performance of FEC codes in coded communications communications systems, especially to very low error rates [33, 34].

In this chapter, we discuss the use of a *dual adaptive importance-sampling* (DAIS) technique based on two uses of the MMC technique [4–7, 31] in order to obtain the performance curves for RCD codes under SPA decoding. Its advantage and uniqueness is that we can accurately compute, via dual complementary IS Metropolis random-walk simulations [57], the probability of occurrence of coded communication system events, such as received or decoder error, down to extremely low values. Thus, the DAIS technique represents an effort to remedy the inadequacy of the MMC technique by conducting dual complementary MMC simulations and combining their results to estimate $P(E)$.

B. Why DAIS?

The RCD codes are regular LDPC codes with known minimum distance of $d_{\min} = 6$. It is also not very difficult to explicitly obtain the codewords at the minimum distance. Thus, in principle, one could use any one of the techniques mentioned in the previous section, (i.e., the techniques of Sadowsky [66], Ferrari and Bellini [22], or Xia and Ryan [77]) to evaluate RCD codes. Yet, we choose the DAIS technique for this purpose.

One of the chief reasons for using DAIS to evaluate RCD code performance is that it is an indigenous technique developed for the purpose of general FEC code evaluation to low error-rates, which makes it easily accessible for performing simulations. The evaluation of RCD codes via DAIS also serves the complementary purpose of validating the DAIS algorithm. In fact, the observations from evaluating RCD codes have provided us with useful pointers for improving DAIS itself. As we will see in Section V.E, use of DAIS also allows us to explain the behavior of the RCD code performance curves with respect to the code union bound in terms of the relative contribution of decoder failures and decoding to wrong codewords to the overall error probability. Finally, DAIS has proven to be capable of estimating lower WERs than those demonstrated by the technique of Xia and Ryan [77] for the $n = 96$, $k = 50$ code, which is discussed in Section V.D.

In the next section, we describe the DAIS technique based on two uses of the MMC algorithm. Here, we also note that since its conception, the DAIS technique has been under further development and refinement at UMBC, and patent and copyright applications have been submitted. Efforts to commercialize this product via licensing to researchers and communications software companies are also in progress.

C. The Dual Adaptive Importance Sampling (DAIS) Technique

The central idea in applying the DAIS technique to coded communications is to perform the MMC technique twice, under *unconstrained* and *constrained* assumptions on the sample space of the random variable under consideration, then use and combine their results to accurately estimate the probability of the desired event [33].

We next describe the important aspects of the DAIS technique in a general framework that makes minimal assumptions on the nature of the problem or the event whose probability is desired to be estimated.

Let $\mathbf{z}$ represent the multi-dimensional random variable (noise vector) of the underlying sample space (Γ) for the problem with known *pdf* $\rho(\mathbf{z})$. Let E denote the event whose probability is to be estimated. The occurrence of E is a function of the value of $\mathbf{z}$, and perhaps, other factors. Importantly, specific regions in sample space where E occurs are unknown a priori.

A scalar mapping $V = f(\mathbf{z})$ from the multi-dimensional sample space to a single-dimensional space (such as the real line or its subset) is chosen that is problem-specific and may often represent some form of distance measure on the sample space (such as the l_2-norm [25]) or modifications of such measures. A desirable mapping represents an ordering of easy-to-define regions in the sample space based on the relative probability of occurrence of event E. We call V the control variable for the simulation problem.

Let Γ_V be the significant range of V such that values of V not in Γ_V have a negligible contribution to $P(E)$, the probability of occurrence of E. We then partition the region Γ_V into a large number (M) of equal-width bins.

In the first use of the MMC algorithm, which defines the unconstrained simulation, the optimal biased *pdf* for $\mathbf{z}$, i.e., $\rho^j(\mathbf{z}) \to \rho^*(\mathbf{z})$ is iteratively estimated (j denotes the iteration number). In this case, the Metropolis random-walk at each iteration is not constrained to specific regions of sample space. Then $\rho^*(\mathbf{z})$

is used to perform IS simulation using an unconstrained Metropolis random walk to obtain estimates of the *pdfs* $p(V)$ and $p(V, E)$.

In the second use of the MMC algorithm, which defines the constrained simulation, we iteratively estimate the optimal biased *pdf* for $\mathbf{z}$ given E, i.e., $\rho^j(\mathbf{z}|E) \to \rho^*(\mathbf{z}|E)$. In this case, the Metropolis random-walk at each iteration is constrained to regions of sample space (and equivalently regions of Γ_V) where event E is guaranteed to occur. Then $\rho^*(\mathbf{z}|E)$ is used to perform IS simulation using the constrained Metropolis algorithm to obtain an estimate of the *pdf* $p(V|E)$.

The results of the constrained and unconstrained MMC simulations are then combined as follows: we scale the estimate of $p(V|E)$, obtained from the constrained simulation, to fit that of $p(V, E)$, obtained from the unconstrained simulation, over the range of values of V where both estimates are sufficiently reliable. Note that this scaling range is problem-specific and usually corresponds to the control variable bins in which the number of recorded error events exceeds a specified threshold. A flowchart of the DAIS technique is shown in Figure V.1.

In the two sections that follow, we apply the DAIS technique to evaluate performance of LDPC codes (including RCD codes) on a noisy communications channel with binary phase-shift-keying (BPSK) modulation and additive white Gaussian noise (AWGN). The LDPC codes are decoded via the sum-product algorithm (SPA) [41] employing the log-likelihood modification of Futaki and Ohtsuki [26]. The maximum number of SPA iterations is set to 50.

Given the channel model, the noise vector $\mathbf{z} = (z_1, \ldots, z_n)$ is multivariate Gaussian with joint *pdf* $\rho(\mathbf{z}) = \prod_{l=1}^{n} \rho_l(z_l)$, where n is the code length. We define the control variable as

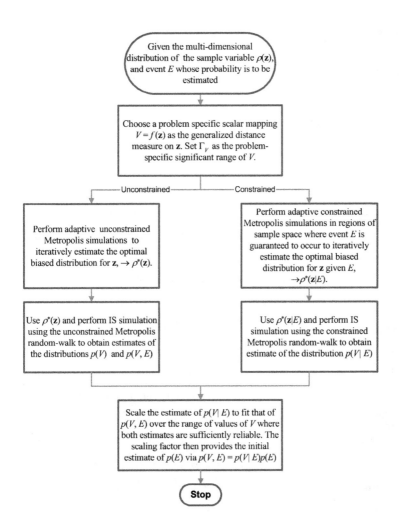

Figure V.1: Flowchart for the dual adaptive importance sampling technique.

$$V(\mathbf{z}) = \left\{ \frac{1}{n} \sum_{l=1}^{n} [H(q_l z_l) z_l]^2 \right\}^{1/2}, \qquad (\text{V.1})$$

where $q_l = (-1)^{b_l}$ with b_l being the transmitted bit in the lth position, and $H(x) = 1$ if $x > 0$ and $H(x) = 0$ otherwise. We constructed $V(\mathbf{z})$ so that a noise component z_l contributes to V only if it may produce a bit-error at the input to the decoder. We say that a received word with a noise realization $\mathbf{z}$ generates an error if the LDPC decoder cannot decode it to the transmitted codeword within 50 iterations.

In our simulations, we set $M = 300$. At the first iteration of the unconstrained simulation, the Metropolis random walk is restricted to $N^{j=1} = 5000$ samples, while for the constrained simulation $N^{j=1} = 10,000$. We increase the number of samples for the Metropolis random walk after each iteration so that $N^{j+1} = 1.3 \, N^j$. In each case, we set $P_k^1 = 1/M$, $k = 1, \dots, M$. The P_k^j is the vector that relates the biased pdf at the jth MMC iteration, $\rho^{*,j}(\mathbf{z})$, to the original pdf $\rho(\mathbf{z})$, for the kth bin via $\rho^{*,j}(\mathbf{z}) = \rho(\mathbf{z})/(c^j P_k^j)$, where k indexes the bin of the control variable to which the noise realization $\mathbf{z}$ maps. The quantities P_k^j satisfy $P_k^j > 0$ and $\sum_{k=1}^{M} P_k^j = 1$, and c^j is an unknown constant that ensures $\int_\Gamma \rho^{*,j}(\mathbf{z}) \, d\mathbf{z} = 1$. We assume the simulation to have sufficiently converged when $\max_k |(P_k^j - P_k^{j+1})/P_k^{j+1}| < 0.1$. This convergence requires $\approx 10^6$ to 10^8 samples in total, with the samples increasing on average with increasing E_b/N_0. Also, in both the constrained and unconstrained simulations, we initialize each MMC iteration with a $\mathbf{z}$ that gives a decoder error.

D. Application of DAIS to Evaluate an $n = 96$, $k = 50$ Regular LDPC Code

In this section, we study the performance of a regular $(96, 50)$ LDPC code with a code rate of $R = 50/96$. The parity-check matrix ($\mathbf{H}$) of this code can be found at [48].

There are multiple reasons for choosing this code to demonstrate DAIS. First, although not an RCD code, it is a regular LDPC code with column-weight $\gamma = 3$. In this regard, this code is similar to RCD codes. Second, this is the same code studied by Xia and Ryan [77]. This allows us to make a valid comparison between DAIS and the importance sampling technique of Xia and Ryan. Finally, it is also possible to exactly compute the first two non-trivial coefficients of the code's weight enumerator function (3 and 24 codewords at Hamming weights 6 and 8, respectively), which allows us to compare our simulation results with the code's union-bound performance [23]. Note that both [77] and [48] refer to this code as a $(96, 48)$ code, but 2 of the 48 rows of $\mathbf{H}$ are linearly dependent.

Figure V.2 shows the different *pdfs* for the DAIS simulation of the $(96, 50)$ code at 11dB. The dashed line shows the *pdf* $p(V)$ from the unconstrained simulation. The dots represent the estimate of $p(V, E)$, while the circles show the number of error events recorded in each bin for the unconstrained simulation. The number of recorded errors rapidly decreases to 0 as V decreases towards 0.4, which is where the contribution of error events to the probability-of-error $(P(E))$ tends to be largest. This is because the unconstrained simulation does not adequately sample the higher-probability smaller-noise realizations that generate errors. As a result, the unconstrained simulation by itself does not provide

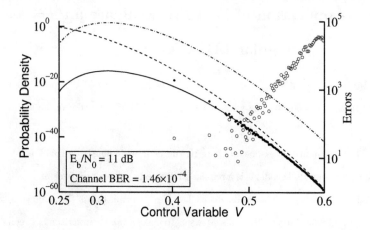

Figure V.2: DAIS results for the $(96, 50)$ LDPC code at 11 dB. Dashed curve: *pdf* of V. Dots: joint *pdf* of V and errors, both from the unconstrained simulation. Dash-dot curve: *pdf* of V conditioned on errors from constrained simulation. Solid curve: joint *pdf* of V and errors obtained by scaling dash-dot curve to fit the dots for $V > 0.55$. Circles: number of decoder errors in the unconstrained simulation.

a good estimate of $P(E)$.

The dash-dot curve in Figure V.2 shows the *pdf* $p(V|E)$ from the constrained simulation, while the solid line shows the new estimate of the *pdf* $P(V, E)$ obtained by scaling $p(V|E)$ to fit the estimate of $p(V, E)$ from the unconstrained simulation for $0.55 < V < 0.6$. This range of V corresponds to those bins where an adequate number of error events are recorded in the unconstrained simulation. That the unconstrained simulation by itself is inadequate to estimate the WER is evidenced by the fact the the estimate of the *pdf* $p(V, E)$ from the unconstrained simulation is not well defined near its peak (compare with the solid curve, which represents the new estimate of the *pdf* $p(V, E)$ after the scaling operation has been performed).

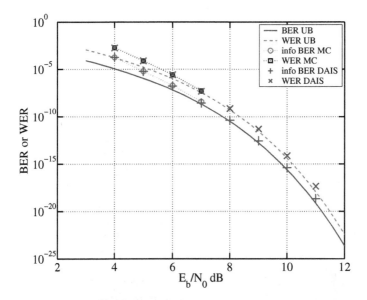

Figure V.3: WER and BER performance curves for the (96, 50) LDPC code. Dotted curve with circles and dotted curve with squares : MC BER and WER estimates, respectively. Plus and cross symbols: DAIS BER and WER, respectively. Solid and dashed curves: BER and WER union bound approximations, respectively, based on codewords at Hamming weight 6 and 8 [23].

Figure V.3 shows the performance curves for the $(96, 50)$ code, the $\times$ and $+$ symbols denote the decoder output BER and WER estimates, respectively, obtained via MC. The dashed curve with □ and dash-dot curve with ○ denote the decoder output BER and WER estimates, respectively, obtained using DAIS. Finally, the solid curve and dotted curve denote the BER and WER union bounds, respectively [23], which can be computed as

$$\mathrm{UB_{WER}} = \sum_{w=d_{\min}}^{n} A_w Q\left(\sqrt{2wr\frac{E_b}{N_0}}\right) \qquad (V.2)$$

and

$$\mathrm{UB_{BER}} = \sum_{w=d_{\min}}^{n} \frac{wA_w}{n} Q\left(\sqrt{2wr\frac{E_b}{N_0}}\right), \qquad (V.3)$$

where n is the code length, r is the code rate, A_w is the number of codewords at Hamming weight w, and $Q(x) = \frac{1}{\sqrt{2\pi}}\int_x^\infty e^{\frac{-t^2}{2}}dt$.

The union bound can be closely approximated at high E_b/N_0 by the contribution of low Hamming weight (6 and 8 in this case) codewords. The SPA for LDPC codes approximates the ML decoder [69]. Hence, we would expect the SPA to perform worse than the union bound on ML decoding at high E_b/N_0. Our results from DAIS are consistent with this expectation and indicate that DAIS can simulate WER and BER performance of codes at very low values. We also observe excellent agreement between the results obtained by DAIS and MC, wherever MC results are available (DAIS falls within the 99% error bars for MC), which further validates DAIS.

Our argument that the true code performance should be close to the union bound at high E_b/N_0 is further bolstered by the observation that for MC simulations, as E_b/N_0 increases, the contribution of the probability of decoding-to-wrong-codewords progressively dominates the WER. For example, at $E_b/N_0 = 4$ dB, 216 of 1888 word errors recorded were due to decoding to wrong codewords (the rest were decoder failures), whereas at $E_b/N_0 = 7$ dB, the corresponding numbers were 40 of 52.

We note that our BER data points do not show a waterfall region since they correspond to large E_b/N_0 relative to the Shannon limit (≈ 0 dB for our code), and since the code is not very long.

A measure of DAIS's gain over MC is given by the ratio of the number of samples (codewords) required to achieve a given WER at a given E_b/N_0; e.g., at $E_b/N_0 = 10$ dB, WER $\approx 10^{-14}$ is obtained by DAIS using 8×10^7 codewords (unconstrained) $+ 3 \times 10^7$ codewords (constrained) $= 11 \times 10^7$ codewords (total), whereas MC would require $\geq 10^{15}$ codewords (assuming ≥ 10 word error events). Thus the gain is $\frac{10^{15}}{11 \times 10^7} \approx 9 \times 10^6$. Our current experience suggest that DAIS's gain increases with decreasing WER, but the accuracy of DAIS as an estimator, and its dependence on the number of codewords or codeword length, is unknown at this time, and a subject of continuing research.

E. RCD Code Performance Evaluations via DAIS

In this section, we present performance curves for different RCD codes on a BPSK-AWGN channel under SPA decoding with a maximum of 50 decoder iterations. The very low error rates are estimated via DAIS, while the higher error rates are estimated via DAIS as well as standard MC simulations. For all RCD codes, the union bound may be easily computed since it is known that the RCD code parameterized by η has $A_6 = \eta\binom{\eta}{3}$ codewords at the minimum distance of 6. As explained previously, the union bound serves as a lower bound on the performance of any sub-optimal decoder including the SPA at high values of E_b/N_0 (The ML decoder is the optimal decoder).

In all the performance curve figures, the solid and dashed curves represent the union bound on the BER and WER, respectively. The + and the × symbols indicate the BER and WER obtained via DAIS. The dotted line with o symbols and the dotted line with □ symbols represent the standard MC BER and WER, respectively. The 99% confidence intervals have been computed for the standard MC WER results based on a Gaussian assumption [36]. The range of the confidence intervals is plotted as vertical bars about the standard MC WER results in the figures.

Figure V.4 shows the performance curves for the $n = 289$, $k = 240$ RCD code with $\eta = 17$, Figure V.5 shows the performance curves for the $n = 529$, $k = 462$ RCD code with $\eta = 23$, while Figure V.6 shows the performance curves for the $n = 1369$, $k = 1260$ RCD code with $\eta = 37$. These specific codes are chosen since they correspond to increasing code lengths in the moderate-to-long range. Longer codes have not been evaluated due to simulation time limitations.

The DAIS simulation results for all three RCD codes are seen to be consistent with the corresponding union bounds (both information BER and WER). For each code, the agreement of the DAIS results with standard Monte Carlo (MC) results is also excellent, with the DAIS WER always falling within the 99% confidence intervals of the corresponding standard MC results. Here, we point

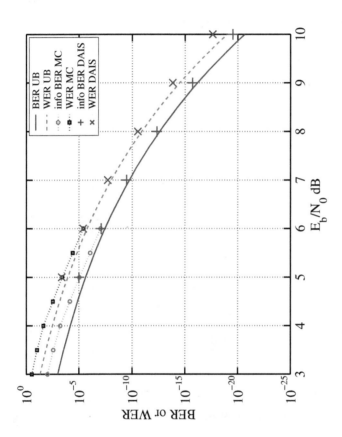

Figure V.4: WER and BER performance curves for the $n = 289$, $k = 240$, $d_{min} = 6$, $\eta = 17$ RCD code under sum-product decoding.

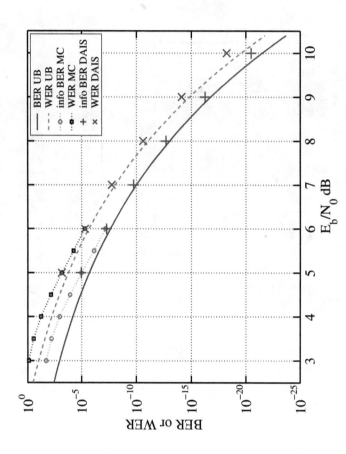

Figure V.5: WER and BER performance curves for the $n = 529$, $k = 462$, $d_{\min} = 6$, $\eta = 23$ RCD code under sum-product decoding.

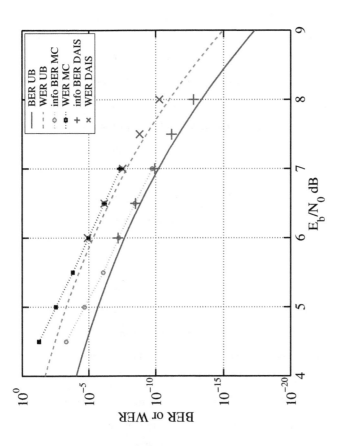

Figure V.6: WER and BER performance curves for the $n = 1369$, $k = 1260$, $d_{min} = 6$, $\eta = 37$ RCD code under sum-product decoding.

out that the 99% confidence intervals are mostly not easily discernible in the plots
on account of the small vertical range that they span. A significant exception is
the standard MC WER at 7dB for the $\eta = 37$ RCD code (see Figure V.6). The
vertical bar of the 99% confidence interval in this case is much larger and clearly
visible on account of the relatively few errors (17 errors in 4×10^8) recorded
during the standard MC simulation. Further, the DAIS WER at this SNR value
is also seen to be within the 99% confidence interval. Consistency of the DAIS
results with the union bound and the standard MC results, strongly hints at the
ability of DAIS to satisfactorily estimate low WER and BER values. However,
the exact accuracy of the DAIS estimator is unknown at this time and needs to
be determined.

We observe that all the three RCD codes considered provide a WER of just
below 10^{-10} at $E_b/N_0 = 8$dB. The shorter code gives better WER performance
at lower E_b/N_0 values, while the longer code is expected to give better WER
performance at higher E_b/N_0 values.

The evaluation of WERs down to 10^{-15} posed hardly any problems for the
$\eta = 17$ and the $\eta = 23$ RCD codes. However, with the $\eta = 37$ RCD code, there
has been considerable difficulty in estimating the WER and BER via DAIS for
$E_b/N_0 \geq 8$dB. The biggest problem has been the inability to record a sufficient
number of error events during the unconstrained simulation at these E_b/N_0 values. The constrained simulation, in contrast, proceeds quite smoothly even at
these high E_b/N_0 values. The problem can be potentially remedied by extending
the right end of Γ_V appropriately, so as to include those bins of V where nearly
all the noise vectors generated lead to decoder error. However, this requires
the generation of a much larger number of noise vectors in order to converge

to the biased *pdf*. Similarly, a much larger number of noise vectors need to be generated and the corresponding received words decoded in order to obtain the results from the unconstrained simulations.

This may, in fact, indicate one of the possible limitations of the DAIS technique in evaluating codes of longer length. In particular, as code length increases, dimensionality of Γ and its partitions that map to bins of V increases. Hence, maintaining a given level of statistical accuracy in sampling each partition of Γ requires more samples for the longer code. Additionally, the time required to perform a single SPA iteration on a received word of a code is linear in the length of the code for fixed column weight [47]. Both these factors suggest that the time required to obtain satisfactory unconstrained simulation results for longer codes is likely to be relatively much higher.

Some modifications to the DAIS technique, such as biasing in the domain of the log-likelihood ratios (LLRs) (see Section V.F.2) or using a different control variable definition may be capable of alleviating the preceding problem and need to be investigated.

It is also interesting to note that the $\eta = 17$ and $\eta = 23$ RCD code performance curves do not exhibit a waterfall region, most likely since the codes are not sufficiently long, and also because of the relatively low minimum distance of the codes, which results in the code performance curve approaching the union bound at a relative high WER of about 10^{-3} to 10^{-4}. However, close observation of the performance curves for the $\eta = 37$ RCD code reveals a weak waterfall region that ends about 5.5dB (WER $\approx 10^{-4}$). The transition from the waterfall region to the *error-floor* region is marked by a perceptible reduction in the steepness of the performance curve. At $E_b/N_0 \geq 5.5$dB, the WER performance curve is seen to be quite close to the union bound, indicating that this region of the performance curve corresponds to the error-floor. Also note that the WER curve for the $\eta = 37$ code diverges more rapidly from the union bound approximation at $E_b/N_0 \leq 5.5\text{dB}$ than for the two shorter RCD codes.

Finally, for all the codes studied (the RCD codes as well as the $(96, 50)$ code), we observe that at low E_b/N_0 (e.g., $\leq 5\text{dB}$ for the $\eta = 23$ code), the WER performance is much higher than the union bound approximation; we refer to this region of the performance curve as the low-SNR region. As the E_b/N_0 is increased, the WER performance comes quite close to the union bound approximation (e.g., $5.5 \leq E_b/N_0 \leq 8\text{dB}$ for the $\eta = 23$ RCD code); we refer to this region of the performance curve as the moderate-SNR region. As the E_b/N_0 is increased further, the WER performance curve is seen to again diverge from the union bound approximation (e.g., $> 8\text{dB}$ for the $\eta = 23$ RCD code); we refer to this region of the performance curve as the high-SNR region. Note that in general the different performance regions span different ranges of E_b/N_0 for different codes.

This behavior can be explained in terms of the relative contribution of decoder failures to the total errors in the control-variable bins that dominate the WER in the different performance regions. As an example, in Figure V.2, the bins of V in the range of 0.28 to 0.35 dominate the WER. These bins are easily determined from the constrained simulation results, since they correspond to the region around the peak of the *pdf* $p(V|E)$. In the low-SNR region, the bins of V that have the most significant contribution to the WER also have decoder failures (as opposed to decoding to wrong codewords) as the dominant component of the errors. An example of WER evaluation in the low-SNR region is shown in Figure V.7, which plots the *pdf* $p(V|E)$ (solid curve) and the partial-*pdf* $p(V, D|E)$ (dotted curve)[1] for the $\eta = 23$ RCD code at 5dB. The event D corresponds to decoding to wrong codewords. Note that the dotted curve is about a factor of 3 below the solid curve where the solid curve reaches its peak. Also, observe that the dotted curve peaks to the left of the solid curve.

In the moderate-SNR region, decoding to wrong codewords contributes significantly to the total error in the bins of V that dominate the WER. Thus, in this performance region, the SPA decoder very closely approximates the ML decoder, resulting in WER performance close to the union bound. An example of WER evaluation in the moderate-SNR region is shown in Figure V.8, which plots $p(V|E)$ (solid curve) and $p(V, D|E)$ (dotted curve) for the $\eta = 23$ RCD code at 7dB. Note that the dotted curve is now only about a factor of 1.5 below the solid curve where the solid curve reaches its peak. Further, both the curves peak at about the same value of V, which is not the case at 5dB.

As we move to the high-SNR regions, the bins of V that have the most significant contribution to the WER again have decoder failures as the dominant component of the errors. Note that as E_b/N_0 changes, the range of values of V that dominates the WER also changes. The direction of change depends on the definition of the control variable. For the control variable defined in equation (V.1), as E_b/N_0 increases the mode of the *pdf* $p(V|E)$ reduces, thus indicating that bins corresponding to smaller values of V dominate the WER.

---

[1]The quantity $p(V, D|E)$ in general does not integrate to unity and, hence, is not strictly a *pdf*. Rather it is one part of a valid 2-dimensional joint *pdf* defined with V as one of the dimensions and a binary random-variable that can either equal D (decoding-to-wrong-codeword) or F (decoder failure) as the other dimension. In the following discussion, we refer to $p(V, D|E)$ as a partial-*pdf*.

An example of WER evaluation in the high-SNR region is shown in Figure V.9, which plots $p(V|E)$ (solid curve) and $p(V, D|E)$ (dotted curve) for the $\eta = 23$ RCD code at 10dB. The dotted curve is now more than a factor of 10 below the solid curve where the solid curve reaches its peak, indicating that the WER contribution due to decoding to wrong codewords is much reduced at this E_b/N_0. Also, in this case, the dotted curve peaks to the right of the solid curve.

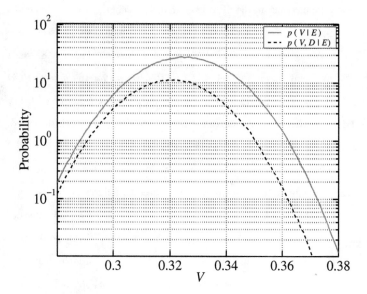

Figure V.7: Plot of the *pdf* $p(V|E)$ and plot of the partial-*pdf* $p(V, D|E)$ for the $\eta = 23$ RCD code at 5 dB.

Lastly we comment on the amount of time taken to obtain some of the DAIS simulation results and how these times compare with those required for standard Monte Carlo. For all codes considered, we have observed that the amount of time

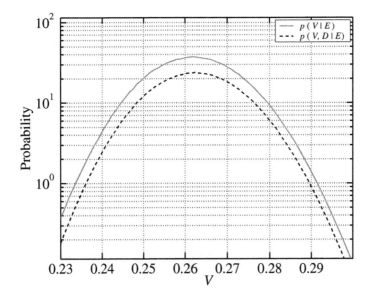

Figure V.8: Plot of the *pdf* $p(V|E)$ and plot of the partial-*pdf* $p(V, D|E)$ for the $\eta = 23$ RCD code at 7 dB.

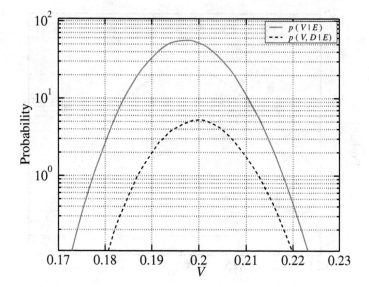

Figure V.9: Plot of the *pdf* $p(V|E)$ and plot of the partial-*pdf* $p(V, D|E)$ for the $\eta = 23$ RCD code at 10 dB.

required by DAIS is much larger than that required by standard Monte Carlo simulations at low values of E_b/N_0. This is primarily because all the bins in the DAIS simulation must be sufficiently populated, irrespective of E_b/N_0, in order to obtain useful results. Also, at low E_b/N_0, the unconstrained MMC simulations are adequate to estimate WER and BER on their own, without any assistance from the constrained simulations. The real advantage of DAIS is revealed only at higher E_b/N_0.

As an example, the standard Monte Carlo evaluation at $E_b/N_0 = 6$dB gave a WER of 3.78×10^{-6} for 271 error events in $\approx 7 \times 10^7$ decoded words, and took 10.37 hours. On the same computer, the total time taken for a DAIS simulation at $E_b/N_0 = 8$dB was 49.1 hours for decoding $\approx 6.5 \times 10^7$ codewords[2], giving a WER of 2.96×10^{-11}. In order to achieve the same WER by standard Monte Carlo with at least 10 errors, one would require to decode $\approx 3.4 \times 10^{11}$ words; this standard Monte Carlo evaluation would require approximately 3 years to perform on the same computer, even if we assume that the average number of SPA iterations per decoding at this E_b/N_0 is half of that at 6dB.[3]

For the $\eta = 37$ code, the standard Monte Carlo evaluation at $E_b/N_0 = 7$dB gave a WER of 4.25×10^{-8} for 17 error events in 4×10^8 decoded words, and took ≈ 340 hours. On the same computer, the total time taken for a DAIS simulation

---

[2]Observe that the time required to decode 6.5×10^7 words in DAIS is much more than the time taken to decode about the same number of words in standard Monte Carlo at 6dB. This discrepancy is easily reconciled by noting that a larger number of error events are generated in the DAIS experiment. Error events usually consume many more SPA iterations on an average per decoding, especially if a large number of decoder failures occur.

[3]We have observed that for standard Monte Carlo, the average number of SPA iterations per decoding decreases with increasing E_b/N_0.

at $E_b/N_0 = 8$dB was 840 hours for decoding $\approx 1.5 \times 10^8$ codewords, giving a WER of 5.5×10^{-11}. In order to achieve the same WER by standard Monte Carlo with at least 10 errors, one would require to decode $\approx 1.8 \times 10^{11}$ words; this would require approximately 10 years to perform on the same computer, even if we again assume that the average number of SPA iterations per decoding at $E_b/N_0 = 8$dB is half of that at 7dB.

F.   Conclusions

1.   Summary and conclusions

We presented performance curves in the form of WER and BER vs. E_b/N_0 for three different RCD codes parameterized by $\eta = 17$, $\eta = 23$, and $\eta = 37$, where the dual adaptive importance sampling (DAIS) technique based on multicanonical Monte Carlo (MMC) simulations was employed to evaluate code performance to low WER and BER. The DAIS technique is employed because it allows us to calculate very low WERs and BERs, far below the capabilities of standard MC estimation techniques. In contrast to standard importance sampling [77], the MMC algorithm, which is a fundamental constituent of the DAIS technique, iteratively approaches the optimal bias without a priori knowledge of how to bias. We improve the WER and BER estimates by combining the results of two MMC simulations in the large noise regions where statistical uncertainty due to sampling from the biased *pdf* is smallest. In the first simulation, which we call the unconstrained simulation, we approximate the probability of decoder errors in the large noise regions. In a second complementary simulation, which we call the constrained simulation, we restrict the MMC random walk to the noise region that produces decoder errors.

Using DAIS, for the RCD code with $\eta = 37$, we were successful in evaluating WERs down to $\approx 10^{-10}$, while for all other codes, we succeeded in estimating WERs to values lower than 10^{-15}. We were not able to estimate WERs lower than 10^{-10} for the $\eta = 37$ RCD code due to the inability to record a sufficient number of error events during the unconstrained simulation at the higher E_b/N_0 values. The simplest way to remedy this problem is to allow more time for the unconstrained simulation, in order to process a larger number of noise vectors. The amount of time required may soon become prohibitive. This may very well represent one of the possible limitations of the DAIS technique in evaluating codes of longer length.

For all codes, the WER and BER results obtained via DAIS were seen to be consistent with standard Monte Carlo simulations and the union bound on maximum-likelihood decoding [23]. In particular, all DAIS WERs fell within the 99% confidence intervals of the corresponding standard MC WERs, wherever both DAIS and standard MC estimates were computed.

The performance of the different RCD codes were compared with each other and a potential limitation of DAIS in evaluating very long codes was discussed. We were also able to explain the relationship between the WER performance curve of the code and the union bound approximation in terms of the relative contribution of decoder failures and decoding to wrong codewords to the total errors in the control-variable bins that dominate the WER at different E_b/N_0. At the low and high E_b/N_0 values, the dominance of decoder failures over decoding to wrong codewords in the region of the control-variable that significantly contributes to the WER leads to a wide separation between the WER performance and the union bound approximation. On the other hand, in the moderate SNR region, the significantly higher error contribution from decoding to wrong codewords, results in the close proximity of the WER performance to the union bound. This also corresponds to the region where the SPA decoder most closely approximates the optimal ML decoder. Rather interestingly, at much higher E_b/N_0, the ability of the SPA to closely approximate the ML decoder diminishes.

The performance curves of the RCD codes are observed to fall off with a rather gentle slope on account of the relatively low value of the RCD code minimum-distance ($d_{\min} = 6$). Also, the $\eta = 17$ and $\eta = 23$ RCD code performance curves do not exhibit a waterfall region, most likely since the codes are not sufficiently long, and also because of the relatively low minimum distance of the codes, which results in the code performance curve approaching the union bound at a relative high WER of about 10^{-3} to 10^{-4}. A weak waterfall region is visible, however, for the $\eta = 37$ RCD code. One would expect the waterfall region to become increasingly prominent only as η increases.

Another interesting conclusion can be drawn by comparing the value of E_b/N_0 that achieves a specified BER for a specific RCD code with the Shannon-limit[4] for the corresponding code rate. We specifically consider the Shannon limit for a binary-input continuous-valued output channel model (which agrees with the BPSK-AWGN model that we have assumed). For such a channel, the capacity is given by equation (3.97) in [3], and the resulting Shannon limits at code-rates of 0.83 ($\eta = 17$ RCD code), 0.87 ($\eta = 23$ RCD code), and 0.92 ($\eta = 37$ RCD code) are 2.31dB, 2.79dB, and 3.53dB, respectively. Comparing these values of E_b/N_0 with the E_b/N_0 values that achieve BER $\approx 10^{-10}$ for the corresponding

RCD codes, which are 7.2dB, 7.1dB, and 7.0dB, respectively, clearly indicates that the gap between this E_b/N_0 and the corresponding Shannon-limit decreases with increasing η. Thus longer RCD codes also have performance closer to their Shannon-limits. The extension of this fact to larger η is further reinforced by our observation that the BER union bounds of the RCD codes when plotted for different η do not show as much horizontal-shift variation (see the curves in Figure V.10). as the Shannon-limits (see the lines in Figure V.10) for the same values of η.

From Figure V.10, we also see that the $\eta = 37$ and the $\eta = 101$ codes achieve the union-bound BER of 10^{-15} at the lowest E_b/N_0 among the codes plotted. If we assume the union bound BER performance to be a good measure of SPA BER performance, we immediately see that the longer RCD code does not necessarily provide the lowest E_b/N_0 at a specific desired BER. Rather, for a desired BER, there is a range of values of η for which the corresponding RCD codes provide the lowest E_b/N_0. For a BER of 10^{-15} the range of η turns out to be approximately between 37 and 101. It is also clear that at a given E_b/N_0, the performance improvement that accrues due to increasing code-rate as η increases is offset by the loss in performance due to a corresponding increase in the term $\frac{A_6}{n}$ (see equation (V.3)), which is proportional to η^2.

Based on the performance curves of the three RCD codes, and intuitively, one would expect that for SPA decoding decoder failures will progressively become the more dominant contributor to the probability-of-error over a larger range of E_b/N_0 values as η increases. An indication in this regard is obtained by observing

---

[4]The Shannon-limit is the minimum E_b/N_0 required to transmit at a particular information transfer rate over a specific channel in order to achieve arbitrarily low probability of bit error.

that the SPA WER performance is close to the corresponding union bound over a larger range of E_b/N_0 values for the $\eta = 17$ RCD code than for the $\eta = 37$ RCD code.

Finally, it is worth pointing out that the decoding complexity of the SPA per iteration is proportional to $6n\gamma$ [47], where n is the code length and γ is the column weight of the regular LDPC code. RCD codes always have $\gamma = 3$. Thus, for a specific code length, we conclude that the SPA decoder for RCD codes has the lowest complexity among decoders for regular LDPC codes of that length[5]. This feature makes RCD codes particularly desirable for high rate low-decoding complexity applications.

2. Future research

1. One of the immediate areas of future work, in the context of the performance curves presented in this chapter, involves the evaluation of longer RCD codes via DAIS or its enhancements. These evaluations will help us further understand the behavior of the performance of the class of RCD codes as a function of increasing η. These evaluations will also help us shed more light on the values of η for which the corresponding RCD codes require the least E_b/N_0 for a desired BER of say 10^{-15}.

2. The evolution of the relative contributions of decoder failures and decoding to wrong codewords, to the WER at specific E_b/N_0 values, as a function of RCD code η, will be helpful in determining how well the SPA approximates the optimal ML decoder at those E_b/N_0 and how this approximation

---

[5]Regular LDPC codes usually always have $\gamma \geq 3$

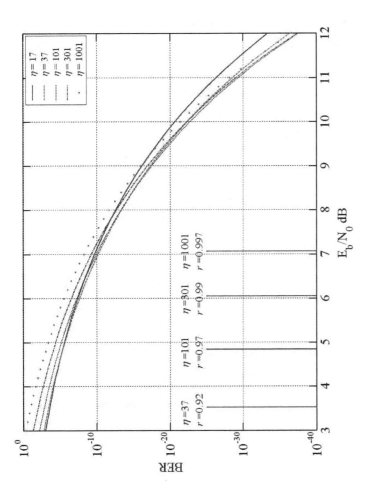

Figure V.10: Plot of the union bound on BER (curves), and Shannon-limit on BER (vertical lines) for RCD codes of various η

evolves as a function of η.

3. The evaluation of the RCD code performance via DAIS has provided us with a lot of insight into the intricacies of the DAIS technique. Certain limitations of the DAIS algorithm also were discovered as a result of the evaluations (see discussion in Section V.E). Some of the other future work that can be envisioned is in the area of developing and improving the DAIS algorithm itself, so that it is capable of evaluating longer codes in lesser time. Next, we list some of the potential modifications that may help improve the DAIS technique.

   (a) The evaluation of the codes presented in this chapter was performed by iteratively computing the biased *pdf* in the space of the noise random variable $\mathbf{z}$. The sum-product algorithm [41] for decoding LDPC codes requires the *log-likelihood ratios* (LLRs) [47] or the *a posteriori probability* (APP) *ratio* of each of the bits in the received word as initialization. Under a BPSK-AWGN channel model the LLRs or the APPs can be easily represented as functions of $\mathbf{z}$. We could also envision an iterative computation of the biased *pdf* in the space of LLRs or APPs, which may prove to be a more advantageous representation.

   (b) The control variable for the LDPC code evaluation was chosen in an ad-hoc fashion. We would like to investigate certain other control variable definitions in order to study their relative performances as well as to determine, if possible, any optimality conditions that the control variable must satisfy.

   (c) In general, the control variable need not be 1-dimensional. Admittedly, a control variable with very large dimensionality is certain to make the DAIS technique intractable. There may be certain advantages, however, in considering 2 or 3-dimensional control variables. For example, consider a 2-dimensional control variable where one of the dimensions is as defined in equation (V.1), while the other dimension corresponds to a binary variable that is an indicator for the event E. In such a case, it may be possible to eliminate the two separate simulations that need to be performed in the DAIS technique.

4. Currently there exist no analytical expressions for the variance of the error probability estimates obtained by the DAIS technique. The task of evaluating the variance of the MMC estimator is quite complicated due to its iterative nature and the fact that the biased *pdf* is sampled using the Metropolis random walk that introduces correlation between successive samples. Some research being currently conducted on the estimator variance of MMC suggests that methods based on bootstrap techniques may provide a solution [43]. The analysis of the DAIS estimator introduces a further level of complexity since it is based on the results of two separate MMC estimates that are non-linearly combined.

Chapter VI.

SNR-INVARIANT IMPORTANCE SAMPLING FOR HARD-DECISION DECODING PERFORMANCE OF LINEAR BLOCK CODES

In this chapter, we present an importance sampling (IS) technique for evaluating the word-error rate (WER) and bit-error rate (BER) performance of binary linear block codes under hard-decision decoding (HDD). This IS technique takes advantage of the invariance of the decoding outcome to the transition probability of the binary symmetric channel (BSC) given a received error-pattern, and is equivalent to the method of stratification for variance reduction. A thorough analysis of the accuracy of the invariance IS (IIS) estimator based on computing its relative bias and variance is provided. Under certain conditions, which may be achieved fairly easily for certain code and decoder combinations, we demonstrate that it is possible to use the proposed IIS technique to accurately evaluate (where accuracy is quantified by the variance and relative bias of the estimator) the

WER and BER to arbitrarily low values. Further, in all cases, the probability estimates obtained via IIS always serve as a lower bound on the true probability values.

The IIS technique was inspired, in part, by the analysis of error-patterns of Hamming weight 3 and 4 for BFA decoding of the class of RCD codes. The idea arose from the observation that the WER performance of RCD codes could be estimated very accurately at high signal-to-noise ratio (SNR), if the exact fraction of error patterns that resulted in an error event was known for the smallest Hamming weight (3 in the case of RCD codes) for which errors were known to occur (see Chapter III). In this chapter, we intentionally evaluate a couple of codes with higher minimum distance than the RCD codes in order to demonstrate the wider applicability of the IIS technique. The IIS technique, of course, has its limitations. These are discussed towards the end of the chapter. Evaluations of a couple of RCD codes using IIS are also provided for purposes of comparison with performance curves presented in Chapter III and some observations on their performance are made.

A. Background and Introduction

The performance of forward error correction (FEC) codes in digital communication systems is almost exclusively studied and compared on the basis of performance curves - plots of bit-error rate (BER) and/or word-error rate (WER) versus the information signal power to noise power ratio (SNR). In general, the exact analytical evaluation of FEC code performance curves under a specific decoding scheme is extremely difficult to perform and computationally intractable, even for moderately long codes. Consequently, one of the preferred methodologies employed by communications engineers and researchers to study code performance has and continues to be based on performing Monte Carlo (MC) computer simulations.

Monte Carlo simulations have been quite effective in obtaining performance curves down to BER values as low as approximately 10^{-7}. Attempting to obtain performance curves for lower values of BER, however, still remains a challenge, even with current computing capabilities because of the inordinate amounts of time required - on the order of months or years - to obtain meaningful results at such low BER values.

One of the most promising areas in the context of evaluating FEC codes to progressively smaller BER values has been the study of importance sampling (IS) techniques [22, 33, 36, 42, 66, 76, 77, 79]. Importance sampling is based on increasing the frequency of low probability error events of interest by sampling these events according to a biased probability density function (*pdf*), which is different from the channel noise *pdf*, and whose choice is FEC code specific.

Most of the proposed techniques for evaluating FEC codes via IS rely on using some code-property to obtain a rough estimate of the region in noise space that is a major contributor to the BER. For example, [22, 66, 76] exploit the knowledge of the codewords at the minimum distance (d_{min}) to determine the biased *pdf* by translating the mean of the original *pdf* towards neighboring codewords, in order to accelerate the production of error events. This knowledge, however, is practically unobtainable for codes in general, especially as the code length increases. One recent technique [79], uses knowledge of the error-correcting capability of codes (a code parameter that may not be known in general) in order to determine the region in the space of error patterns (called the IS region) such that all error patterns outside this region are decoded successfully. The probability of the IS region is computed recursively via combinatorial expressions, and the biased distribution for each Hamming weight is then determined based on the total number of error patterns at that Hamming weight that belong to the IS region. This technique, while restricted to hard-decision decoding, is particularly suited for serially concatenated coding schemes (product codes) where it is possible to easily characterize the error correcting capability of the constituent codes and, hence, the IS region.

A few other techniques [33,77] focus their attention on soft-decision decoding (SDD) of FEC codes. The method of [77], which applies to linear block codes, does away with the need for knowing the codewords at $d_{\min}$ and, instead, uses knowledge of the bipartite graph of the code. The task of finding the biased *pdf* for the entire code, then, is reduced to that of determining the biased *pdf* for constituent single-parity check codes that are defined by each parity-check equation. The dual-adaptive importance sampling technique of [33], which is also presented in Chapter V of this dissertation, estimates the probability of error events by combining dual complementary IS simulations. Each of the two constituent simulations adaptively determines a biased *pdf* using the multicanonical Monte Carlo (MMC) method [7], which is an iterative procedure that requires relatively little a priori knowledge of how to bias.

In this chapter, we present an IS technique for hard-decision decoding of linear block codes for the binary symmetric channel (BSC) that exploits the concept of invariance of a hard-decision FEC decoder output to the BSC transition probability p given a received-word that belongs to the set of all binary tuples of length n. Investigation into this technique was motivated by our observation for RCD codes that the exact fraction of error patterns that resulted in an error event for the smallest Hamming weight (3 in the case of RCD codes) at which errors were known to occur was sufficient to estimate the WER performance of the code over a range of SNR, especially at the high end.

We first explain this concept of invariance in detail. Then, we develop the SNR-invariant IS (IIS) method. Next, we analyze the IIS estimator performance in terms of its bias and variance. We present results from IIS simulations for some chosen codes (RCD and others) of moderate length, and finally end with the summary and conclusions.

A salient feature of our proposed IS technique, which distinguishes it from previous IS techniques for hard-decision decoding, is that it does not require specific knowledge of the codewords at d_{min} to construct the biased *pdf*. Some knowledge of the error correcting capability of the decoder[1] for the code under consideration, however, may be used to enhance or ascertain the validity of the estimates. Our technique is closest to that presented in [79], but is more general in conceptual applicability. In fact, its equivalence to the method of stratification [42][2] makes it one of the most fundamental techniques for variance reduction. In this chapter, we show that under suitable conditions, which are primarily determined by code length and decoder error correcting capability, this technique can accurately estimate code performance to arbitrarily low error-rates.

---

[1]The error correcting capability of a decoder for a particular code may be different from the error correcting capability of the code itself.

[2]This equivalence is established in Section VI.C.

B. Invariance of Decoder Output to SNR

A hard-decision decoder (HDD) for a (n, k) binary linear block code on a BSC accepts a length-n binary word received at the channel output (called the received-word) and attempts to estimate the transmitted codeword. Let $\mathcal{B}^n$ denote the space of all length-n binary words. We define the *capture region*[3] of a codeword C, under a particular HDD, as the set of received-words in $\mathcal{B}^n$ that decode to C.

The HDD codeword estimate is completely dependent on the received-word, i.e., the same received-word always generates the same codeword estimate irrespective of the BSC transition probability p that caused the received-word. Thus, the capture region of any codeword C in $\mathcal{B}^n$ is invariant to changes in p, and hence, the SNR.

It follows that simulations based on a probability distribution on $\mathcal{B}^n$ result in an SNR-invariant characterization of the capture region. Thus, as we will demonstrate, knowledge of the capture region obtained from simulations performed for a particular p may be used for performance evaluation at any other p value. Alternately, simulations may be performed in $\mathcal{B}^n$ without reference to any specific p value, and the results may then be used to evaluate performance over a wide range of p values.

C. SNR-Invariant Importance Sampling

In this section we develop an IS scheme for BER/WER evaluation of HDDs on the BSC with transition probability p (BSC(p)) based on the concept of the

---

[3]The capture region may be called the decoding sphere, although the former term is more accurate since this region may not be spherical

decoder output's invariance to p, i.e., to the SNR.

It is sufficient to transmit the all-zeros codeword for overall code performance evaluation, since the code is linear and the channel is symmetric. We can safely restrict out study to the class of HDDs that satisfy the symmetry conditions of [65] without much loss of generality, since most HDDs of practical interest indeed satisfy the symmetry conditions. These symmetry conditions ensure that the capture region of each codeword is identical when characterized with the corresponding codeword at its center.

In a standard MC simulation to evaluate code performance on a BSC(p), we would transmit the all-zeros codeword and flip (complement) each transmitted bit independently with probability p to obtain the received-word. For small values of p, the received-word, more often than not, is likely to have small Hamming weight. Specifically, the probability of having a received-word at Hamming weight w is $p^w (1-p)^{n-w}$. Unfortunately, a large portion of the simulation effort produces received-words that will decode successfully to the all-zeros codeword: this represents a wasted effort as far as BER/WER performance evaluation is concerned.

Consider the error event E whose probability is to be estimated, and received-words $\mathbf{z} \in \mathcal{B}^n$ drawn from the original pmf $\rho(\mathbf{z})$. A standard MC simulation will attempt to estimate $P(E)$ via

$$P(E) \approx \frac{1}{N} \sum_{i=1,\dots,N} \chi_E(\mathbf{z}^i) , \qquad (\text{VI.1})$$

where $\mathbf{z}^i$ represent the received-words sampled from pmf $\rho(\mathbf{z})$, N is the total number of samples of received-words generated, and $\chi_E(\mathbf{z}^i)$ is the indicator function for E that takes a value of 1 if $\mathbf{z}^i$ causes event E, and a value of 0 otherwise.

In IS, we desire to sample according to a suitable biased *pmf* $\rho^*(\mathbf{z})$ that increases the relative occurrence of received-words that cause E [36, 42]. We then compute $P(E)$ via

$$P(E) \approx \frac{1}{N} \sum_{i=1,\dots,N} \chi_E \left(\mathbf{z}^i\right) \frac{\rho\left(\mathbf{z}^i\right)}{\rho^*\left(\mathbf{z}^i\right)} , \qquad (\text{VI.2})$$

where $\mathbf{z}^i$ now represent the received vectors sampled from the *pmf* $\rho^*(\mathbf{z})$.

In choosing a biased *pmf* for IS, we desire to sample only from a set of received-words (error-patterns) with Hamming weights within a predetermined range. Let w be the Hamming weight of a received-word. Then a reasonable range for w may be given by $1 \leq w \leq w_{\max} = \lceil \frac{d''}{2} \rceil + c$, where d'' is the smallest number for which it is known that codewords of Hamming weight d'' exist, and c is a small positive integer. For each $w \in \{1, \dots, w_{\max}\}$, we generate $N_w \leq \binom{n}{w}$ codeword samples. The specific values of N_w are determined by practical considerations such as allowable simulation time and desired error rate. They may also be obtained via a constrained optimization [42] based on the requirement to minimize the estimator variance given a fixed total number of samples $N = \sum_{j=1}^{w_{\max}} N_j$ that must be decoded.

Let $wt(\cdot)$ denote the Hamming weight operator. The probability of sampling a weight-w received-word according to this scheme is given by

$$P\{wt(\mathbf{z}) = w\} = \frac{N_w}{\sum_{j=1}^{w_{\max}} N_j} = \frac{N_w}{N} . \qquad (\text{VI.3})$$

For a given weight w, we generate the received-word by sampling from the uniform *pmf* over all possible length-n weight-w binary words. Thus, the probability of choosing a particular weight-w received-word conditioned on knowledge of w is given by

$$P\{\mathbf{z}|wt(\mathbf{z}) = w\} = \frac{1}{\binom{n}{w}}, \tag{VI.4}$$

and the probability of choosing a received-word $\mathbf{z}$ according to this scheme is given by

$$P\{\mathbf{z}, wt(\mathbf{z}) = w\} = P\{\mathbf{z}|wt(\mathbf{z}) = w\} \cdot P\{wt(\mathbf{z}) = w\} \tag{VI.5}$$

$$= \frac{1}{\binom{n}{w}} \cdot \frac{N_w}{\sum_{j=1}^{w_{\max}} N_j} = \frac{1}{\binom{n}{w}} \cdot \frac{N_w}{N}. \tag{VI.6}$$

Equation (VI.6) defines the biased *pmf* $\rho^*(\mathbf{z})$ that is used to generate the received-words for IIS. Further, the probability of choosing a specific received-word according to the original *pmf* is

$$\rho(\mathbf{z}) = p^{wt(\mathbf{z})}(1-p)^{n-wt(\mathbf{z})} . \tag{VI.7}$$

We can now perform IS simulations to compute $P(E)$ via equations (VI.2), (VI.6), and (VI.7), i.e.,

$$P(E) \approx \frac{1}{N} \sum_{j=1,\dots,N} \chi_E\left(\mathbf{z}^j\right) p^{wt\left(\mathbf{z}^j\right)} (1-p)^{n-wt\left(\mathbf{z}^j\right)} \cdot \frac{\binom{n}{wt(\mathbf{z}^j)} \cdot N}{N_{wt(\mathbf{z}^j)}} ,$$

$$\text{(VI.8)}$$

$$\approx \sum_{j=1,\dots,N} \chi_E\left(\mathbf{z}^j\right) \frac{p^{wt\left(\mathbf{z}^j\right)} (1-p)^{n-wt\left(\mathbf{z}^j\right)} \cdot \binom{n}{wt(\mathbf{z}^j)}}{N_{wt(\mathbf{z}^j)}} . \qquad \text{(VI.9)}$$

By grouping the received-words at each Hamming weight, equation (VI.9) can be rewritten as

$$P(E) \approx \sum_{w=1}^{w_{\max}} p^w (1-p)^{n-w} \binom{n}{w} \cdot \frac{E_w}{N_w} , \qquad \text{(VI.10)}$$

where E_w is the number of word-error events recorded for Hamming weight w when N_w received-words of Hamming weight w are decoded. We refer to the fraction $\frac{E_w}{N_w}$ as the raw WER at Hamming weight w, while the wth term in the summation in equation (VI.10) is called the weighted WER at Hamming weight w. Clearly, the raw WER is independent of p, and the influence of p on the probability of event E $(P(E))$ at any SNR is only via the weighting factor $p^w (1-p)^{n-w} \binom{n}{w}$. Figure 1 shows a flowchart of the proposed IIS scheme.

From equation (VI.10), our IS technique is readily seen to be equivalent to the *method of stratification* [42] – a variance reduction technique well known in statistics – with the entire space of received-words partitioned (or stratified) into subsets based on Hamming weight. The theory developed for the method of stratification can then be used to evaluate the IIS estimator mean and variance. Note that unless $w_{\max} = n$ is satisfied, the estimator of equation (VI.10) will be biased. In practice, however, it is desirable to set $w_{\max}$ to a much smaller

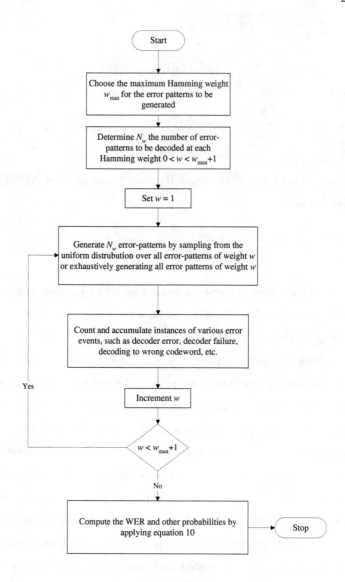

Figure VI.1: Flowchart indicating the major steps in the proposed IIS technique.

number since the contribution of the raw WER at the higher Hamming weights to the total WER is negligible, especially at high SNR.

D. Analysis of the IIS Estimator

The estimator bias and variance (or standard-deviation) are two measures that quantify its accuracy. Estimators having negligible or no bias, and low standard-deviation relative to the estimates obtained, are considered *accurate*: by *accurate*, we mean the estimator has a standard-deviation smaller than the estimated value and negligible relative bias.[4]

We now analyze the IIS estimator accuracy by computing its bias and variance (and, hence, its standard-deviation). Let $\hat{P}(E)$ denote the estimate of the probability $P(E)$, where event $E \triangleq \{\mathbf{z} \in \mathcal{B}^n : \mathcal{D}(\mathbf{z}) \neq \mathbf{0}\}$, $\mathcal{D}(\cdot)$ denotes the decoder function, and $\mathbf{0}$ the all-zeros codeword. Then, from equation (VI.10)

$$\hat{P}(E) = \sum_{w=1}^{w_{\max}} p^w (1-p)^{n-w} \binom{n}{w} \cdot \frac{1}{N_w} \sum_{i=1}^{N_w} \chi_{E|w}\left(\mathbf{z}^i\right) , \qquad \text{(VI.11)}$$

where $\chi_{E|w}\left(\mathbf{z}^i\right)$ is the indicator function for the event E given that a received-word of Hamming weight w was transmitted. Equation (VI.11) can be restated as

$$\hat{P}(E) = \sum_{w=1}^{w_{\max}} P_w \cdot \hat{P}\left(E|w\right), \qquad \text{(VI.12)}$$

where $P_w = p^w (1-p)^{n-w} \binom{n}{w}$ and $\hat{P}(E|w) = \frac{1}{N_w} \sum_{i=1}^{N_w} \chi_{E|w}\left(\mathbf{z}^i\right)$ is the estimator of the probability of event E conditioned on the received-word having Hamming

---

[4]This term is defined later in this section

weight w. Finally, the true probability of event E, $P(E)$, can be written as

$$P(E) = \sum_{w=1}^{n} P_w \cdot P(E|w),\qquad\text{(VI.13)}$$

where $P(E|w)$ is the true probability of event E conditioned on the received-word having Hamming weight w.

The mean of each of the estimators $\hat{P}(E|w), w = 1, \ldots, w_{\max}$, can be computed by

$$\mathbb{E}\{\hat{P}(E|w)\} = \frac{1}{N_w} \sum_{i=1}^{N_w} \mathbb{E}\{\chi_{E|w}(\mathbf{z}^i)\} = P(E|w),\qquad\text{(VI.14)}$$

where $\mathbb{E}\{\cdot\}$ represents the expectation operator. The above follows since $\chi_{E|w}(\mathbf{z})$ is Bernoulli distributed with $Pr\{\chi_{E|w}(\mathbf{z}) = 1\} = P(E|w)$ and $Pr\{\chi_{E|w}(\mathbf{z}) = 0\} = 1 - P(E|w)$.

From equations (VI.12) and (VI.14), and with the expectation with respect to $\mathbf{z}$ we can readily obtain

$$\mathbb{E}\{\hat{P}(E)\} = \sum_{w=1}^{w_{\max}} P_w \cdot \mathbb{E}\{\hat{P}(E|w)\} = \sum_{w=1}^{w_{\max}} P_w \cdot P(E|w).\qquad\text{(VI.15)}$$

The absolute bias of the estimator is then given as

$$\left|B\left(\hat{P}(E)\right)\right| = |\mathbb{E}\{\hat{P}(E)\} - P(E)| = \sum_{w_{\max}+1}^{n} P_w \cdot P(E|w),\qquad\text{(VI.16)}$$

and can be upper bounded by

$$\left|B\left(\hat{P}(E)\right)\right| \le \sum_{w_{\max}+1}^{n} P_w.\qquad\text{(VI.17)}$$

A quantity that is probably more interesting than the absolute bias is the relative bias $B_r\left(\hat{P}(E)\right)$, which we define as the ratio of the absolute bias to the estimated probability of error:

$$B_r\left(\hat{P}\left(E\right)\right) = \frac{\left|B\left(\hat{P}\left(E\right)\right)\right|}{\hat{P}\left(E\right)} . \tag{VI.18}$$

From equations (VI.17) and (VI.18) it follows that

$$B_r\left(\hat{P}\left(E\right)\right) \leq \frac{\sum_{w_{\max}+1}^{n} P_w}{\sum_{w=1}^{w_{\max}} P_w \cdot \hat{P}\left(E|w\right)} . \tag{VI.19}$$

We observe that the upper bound on the relative bias does not depend on the true values $P\left(E|w\right), w = 1, \ldots, w_{max}$, but only on the corresponding estimates (the relative bias was intentionally defined in this way to facilitate computation). Thus, the bound on the relative bias can be computed based on the IIS simulation results (the quantities P_w are known).

Our aim is to ensure that the IIS estimator relative bias is below a prescribed threshold (say 10^{-3}). Given a choice of $w_{\max}$, we can obtain a range of E_b/N_0 values for which the relative bias is below the prescribed threshold. If the desired threshold is not met, we must increase $w_{\max}$ appropriately. Thus, a trial-and-error approach is required to arrive at an appropriate $w_{\max}$.

An empirical rule-of-thumb does exist to determine $w_{\max}$, however, which is based on the following. At high SNRs, $P\left(E\right)$ is dominated by errors occurring at the smallest Hamming weight. Hence, choosing $w_{\max} = \lceil \frac{d''}{2} \rceil + c$, where d'' is the smallest number for which it is known that codewords of Hamming weight d'' exist, and c is a small positive integer, more often than not, ensures a negligible bias at high SNRs. In addition, the plot of the weighted WER at a given SNR versus Hamming weight serves as a visual indicator of whether or not the IIS estimator bias is negligible. If the weighted WER is decreasing as w increases, for values of w just smaller than $w_{\max}$, and the weighted WER at $w_{\max}$ is much smaller than the maximum of the weighted WER over considered values of w (see Figures VI.2, VI.4, VI.6, and VI.8 for examples), then we are very likely to have a negligible relative bias.

Finally, we note that the IIS estimator tends to underestimate the probability of E, i.e., $\hat{P}(E) \leq P(E)$. Thus, the relative bias computed with respect to $P(E)$ is likely to be even smaller than the relative bias with respect to $\hat{P}(E)$ as defined by equation (VI.18).

Having computed the IIS estimator mean and bias (and relative bias), the IIS estimator variance can be computed as

$$var\{\hat{P}(E)\} = var\{\sum_{w=1}^{w_{max}} P_w \cdot \hat{P}(E|w)\} = \sum_{w=1}^{w_{max}} P_w^2 \cdot var\{\hat{P}(E|w)\} , \quad \text{(VI.20)}$$

which follows from the fact that the estimates $\hat{P}(E|w)$, $w = 1, \ldots, w_{max}$, are computed independently. Since $\chi_{E|w}(\mathbf{z})$ are independent Bernoulli distributed, we have for each w

$$var\{\chi_{E|w}(\mathbf{z}^i)\} = P(E|w)(1 - P(E|w)), \, \forall i , \quad \text{(VI.21)}$$

and

$$
\begin{aligned}
var\{\hat{P}(E|w)\} &= var\{\frac{1}{N_w} \sum_{i=1}^{N_w} \chi_{E|w}(\mathbf{z}^i)\} = \frac{1}{N_w^2} \sum_{i=1}^{N_w} var\{\chi_{E|w}(\mathbf{z}^i)\} \\
&= \frac{P(E|w)(1 - P(E|w))}{N_w} .
\end{aligned}
\quad \text{(VI.22)}
$$

Finally, using equation (VI.22) in (VI.20) results in

$$var\{\hat{P}(E)\} = \sum_{w=1}^{w_{max}} P_w^2 \cdot \frac{P(E|w)(1 - P(E|w))}{N_w} . \quad \text{(VI.23)}$$

We note that it is possible to upper bound the variance of $\hat{P}(E|w)$ via

$$var\{\hat{P}(E|w)\} \leq \frac{P(E|w)}{N_w} , \qquad \text{(VI.24)}$$

which leads to the following inequality

$$var\{\hat{P}(E)\} \leq \sum_{w=1}^{w_{\max}} P_w^2 \cdot \frac{P(E|w)}{N_w} . \qquad \text{(VI.25)}$$

Unfortunately, equation (VI.23) cannot be used directly to compute the estimator variance since $P(E|w)$ is unknown. One alternative is to approximate $P(E|w)$ by $\hat{P}(E|w)$. Another is to use the raw simulation results at Hamming weight w to estimate the variance of $\chi_{E|w}(\cdot)$ via the unbiased sample variance estimator. Both the preceding techniques will work so long as we have non-zero estimates of $\hat{P}(E|w)$ for all values of w, where it is known that some weight-w error-patterns result in errors. Often times this requirement may not be satisfied. In particular, there may exist a range of Hamming weights $w' \leq w \leq w^*$ for which $\hat{P}(E|w) = 0$ based on N_w received-words, but it is not known for certain that $P(E|w) = 0$. The Hamming weight w^* may then be defined as the largest Hamming weight such that transmission of N_{w^*} codewords does not result in a single decoder error and it is not known that $P(E|w^*) = 0$. A similar definition also can be constructed for Hamming weight w'. Here, we note that for small non-zero Hamming weights w, it is possible to ascertain if $P(E|w) = 0$ by exhaustively generating and decoding each of the $\binom{n}{w}$ error-patterns at that Hamming weight. It may also be possible to exploit certain theoretical properties of the decoder to rule out the possibility of error patterns below a certain Hamming weight resulting in decoder error.

It is possible to upper bound $P(E|w)$ when $w' \leq w \leq w^*$; such upper bounds, however, cannot be claimed with complete certainty. Rather, there always exists a confidence ($< 100\%$) associated with the bound. The mechanism to compute the upper bound on $P(E|w)$ and the associated confidence is founded on the concept of computing confidence intervals [58] based on simulation estimates. For this particular case, it can be shown that when $\hat{P}(E|w) = 0$,

$$Pr\{P(E|w) \geq \frac{\beta}{N_w}\} = \left(1 - \frac{\beta}{N_w}\right)^{N_w} \approx e^{-\beta} , \qquad (VI.26)$$

where N_w is the number (assumed large) of error-patterns decoded at weight w and β is a small integer. Equation (VI.26) leads to the following bound:

$$Pr\{P(E|w) < \frac{\beta}{N_w}\} = 1 - \left(1 - \frac{\beta}{N_w}\right)^{N_w} \approx 1 - e^{-\beta} , \qquad (VI.27)$$

where $(1 - e^{-\beta}) \times 100\%$ is the associated confidence. Thus, $\frac{\beta}{N_w}$ may be substituted as an upper bound for $P(E|w)$, $w' \leq w \leq w^*$, in equation (VI.25), bearing in mind that the associated confidence is $(1 - e^{-\beta}) \times 100\%$. Wherever applicable, we use $\beta = 5$, since this results in a confidence of $> 99\%$. Similarly, for $w > w^*$, $\hat{P}(E|w)$ may be employed instead of $P(E|w)$ to get an approximate upper bound on the variance of $\hat{P}(E)$ as

$$var\{\hat{P}(E)\} \lesssim \sum_{w=w'}^{w^*} P_w^2 \cdot \frac{\beta}{N_w^2} + \sum_{w=w^*+1}^{w_{\max}} P_w^2 \cdot \frac{\hat{P}(E|w)}{N_w} . \qquad (VI.28)$$

Finally the standard-deviation of the estimator may be computed as

$$std.dev.\{\hat{P}(E)\} \lesssim \sqrt{\sum_{w=w'}^{w^*} P_w^2 \cdot \frac{\beta}{N_w^2} + \sum_{w=w^*+1}^{w_{\max}} P_w^2 \cdot \frac{\hat{P}(E|w)}{N_w}} . \qquad (VI.29)$$

We will use equations (VI.19) and (VI.28) or (VI.29) to quantify and ascertain the accuracy of the estimates of the WER for the examples presented in Section VI.E.

E.   Example Application and Results

We present IIS simulation results for the $m = 2, s = 4, \mu = 0$ Euclidean geometry (EG) code with length $n = 255$ and dimensionality $k = 175$ [44], the 4-dimensional (4D) single-parity-check product (SPCP) code of length $n = 4^4 = 256$ and dimensionality $k = 3^4 = 81$ [12], the $\eta = 17$ $(289, 240)$ RCD code, and the $\eta = 37$ $(1369, 1260)$ RCD code. All codes are decoded via bit-flipping decoding (BFD) [38, 52] with a maximum of 20 decoder iterations. A detailed description of the BFD has been previously provided in Subsection III.A.2.

1.   (255,175) EG code

Consider the $n = 255$, $k = 175$ EG code. Table VI.1 lists the word error events and total words transmitted at each of the Hamming weights considered. Zero words were decoded for all Hamming weights not in the table. Clearly, we have $w_{max} = 34$. Figure VI.2 shows the raw WER for the different Hamming weights, and the corresponding weighted WERs at 3, 6, 8, and 9dB.

We see that the raw WER is monotonically non-decreasing with w and is very close to unity when $w = 16$. At the smallest SNR of 3dB, the relative bias upper bound computed via equation (VI.19) gives $B_r(\hat{P}(E)) \leq \frac{4.63 \times 10^{-8}}{4.56 \times 10^{-1}} = 1.02 \times 10^{-7}$, which can be deemed negligible. Further, the relative bias decreases with increasing SNR (see Table VI.2), which is expected since from Figure VI.2

Table VI.1.

Total error patterns transmitted and corresponding word error events recorded at the considered Hamming weights for the $n = 255$, $k = 175$ EG code.

Hamming weight, w	1	2	3-8	9	10	11	12
N_w	255	$\binom{255}{2}$	10^5	10^5	10^5	10^5	10^5
E_w	0	0	0	38	609	6436	28426

Hamming weight, w	13	14	15	16	17	18	19
N_w	10^5	10^5	10^5	10^5	10^5	10^5	10^5
E_w	57900	84794	96129	99298	99929	99995	10^5

Hamming weight, w	20-24	25-34
N_w	10^5	10^4
E_w	10^5	10^4

the weighted WER decreases more rapidly with increasing w near w_{max} as the SNR is increased. Thus, for the SNR values considered, the choice of $w_{max} = 34$ leads to negligible bias, thereby justifying our choice.

Without further knowledge of the code and decoder, more codewords at $w \leq 8$ must be decoded to determine if any errors are possible. In fact, to be absolutely sure, we should exhaustively generate and decode all possible $\binom{n}{w}$ weight-w received-words. This task rapidly becomes prohibitive with increasing n and w and represents a fundamental limitation of this method. This limitation may be less restrictive, however, if more knowledge on the decoder's error-correcting capability is available.

The considered EG code has a parity-check matrix $\mathbf{H}$ with 16 ones in each row and column, with the set of parity-check equations (PCEs) orthogonal or concurrent on every bit-position [44]. As shown in [44], the majority-logic decoder (MLD) is capable of correctly decoding all received-words with $w \leq 8$ for this code. BFD, as an iterative modification of the MLD, also can be also shown capable of correctly decoding all received-words with $w \leq 8$ (see Lemma III.1 in Chapter III). In addition, however, the BFD also can correctly decode quite a large fraction of received-words at some of the immediately higher Hamming weights.

Given this additional knowledge, we may skip the search for error events with received-words of $w \leq 8$. By having recorded a significant error-count for $w \geq 9$, and by knowing that received-words of $w \leq 8$ do not contribute to error events (i.e., $P(E|w) = 0, w = 0, \ldots, 8$), we can now evaluate the estimator variance and standard-deviation in order to characterize its accuracy for SNRs where error events from received-words in the range $w_{min} \leq w \leq w_{max}$ contribute almost entirely to the WER.

The first term on the right of equation (VI.28) vanishes for the case under consideration since for all w for which $\hat{P}(E|w) = 0$, it is known with certainty that $P(E|w) = 0$ (i.e., the set $\{w', \ldots, w^*\}$ is empty). Thus, the estimator variance satisfies

$$var\{\hat{P}(E)\} \lesssim \sum_{w=9}^{w_{\max}} P_w^2 \cdot \frac{\hat{P}(E|w)}{N_w} . \qquad \text{(VI.30)}$$

The standard-deviation computed by evaluating the square-root of equation (VI.30) for various SNR values are given in Table VI.2.

Table VI.2
The WER estimate, upper bound on relative bias via equation (VI.19), and approximate estimator standard-deviation via equation (VI.30) for IIS for different values of E_b/N_0, and standard-deviation of the MC estimator for $N = 2.33 \times 10^6$ assuming true WER is as determined by the IIS estimator. All results for the $(255, 175)$ EG code.

E_b/N_0	3	6	8	9
IIS WER estimate	4.5×10^{-1}	7.3×10^{-6}	4.25×10^{-13}	6.05×10^{-18}
IIS rel. bias upper bound	1.0×10^{-7}	9×10^{-24}	6×10^{-43}	1.5×10^{-56}
IIS estimator std. dev.	1.76×10^{-4}	1.41×10^{-9}	8.6×10^{-19}	9.23×10^{-26}
Std. MC estimator std. dev.	3.26×10^{-4}	1.76×10^{-6}	4.27×10^{-10}	1.61×10^{-12}
(IIS WER assumed true)				

We see that, for the E_b/N_0 values considered, the IIS estimator standard-deviation decreases as E_b/N_0 increases for the same values of N_w and $\hat{P}(E|w)$. Also, relative to the estimated WER value, the IIS estimator standard-deviation is a few orders of magnitude lower for all E_b/N_0 considered, indicating sufficient estimation accuracy.

228

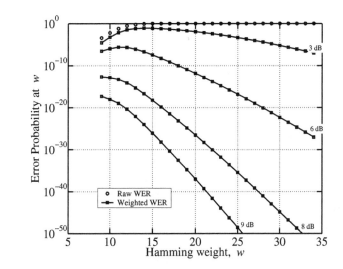

Figure VI.2: Weighted WER at various E_b/N_0 and raw WER vs. Hamming weight w for the $(255, 175)$ EG code under BFD

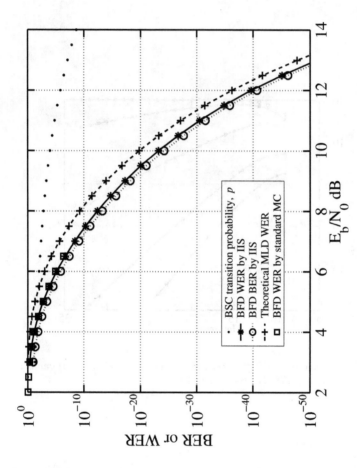

Figure VI.3: Error Probabilities as a function of E_b/N_0 for the (255, 175) EG code

We can now compare the IIS estimator standard-deviation with that of a standard MC estimator for the same total number of decoded received-words ($N = \sum_{w=1}^{w_{\max}} N_w = 2.33 \times 10^6$). For the standard MC estimator based on decoding N received-words, the estimator variance is

$$var\{\hat{P}_{MC}(E)\} = \frac{P\left(E\right)\left(1 - P\left(E\right)\right)}{N} \approx \frac{P\left(E\right)}{N} . \qquad \text{(VI.31)}$$

If we assume that the true probability of error equals that determined by the IIS estimator, we can compute the standard-deviation of the standard MC estimator for $N = 2.33 \times 10^6$ decoded received-words via the square-root of equation (VI.31). These values are listed in the last row of Table VI.2, where we see that, for all E_b/N_0 values considered, the IIS estimator gives lower standard-deviation as compared to the standard MC estimator, with the gap between the two standard-deviations increasing with increasing E_b/N_0. We also note that at 8 and 9dB the standard-deviation of the standard MC estimator exceeds the IIS-estimated WER, indicating that the standard MC estimator is ineffective at estimating these low WERs. Finally, we also point out that the lowest WER that can be estimated by standard MC, using $N = 2.33 \times 10^6$ received-words and adequate reliability[5], is $\approx \frac{10}{2.33 \times 10^6} = 4.3 \times 10^{-6}$. Estimating lower WERs by standard MC would require many more received-words to be decoded.

Figure VI.3 shows plots of various error rates as a function of E_b/N_0 in dB, where the BSC(p) corresponds to a binary phase-shift-keying (BPSK) additive white Gaussian noise (AWGN) channel with symmetric signal levels at ±1. The AWGN variance is σ^2 and $E_b/N_0 = 10\log_{10}\frac{1}{2r\sigma^2}$dB, where $r = k/n$ is the code

[5]In this case, we consider the standard MC estimate to be adequately reliable if ≥ 10 error-events are recorded.

rate. The dots represent $p = Q\left(\frac{1}{\sigma}\right)$, where $Q(x) = \frac{1}{\sqrt{2\pi}} \int_x^\infty e^{\frac{-t^2}{2}} dt$. The dashed line with + symbols shows the MLD WER based on successful decoding of all received-words of $w \le 8$, with the assumption that no received-word with $w \ge 9$ is decoded successfully. The MLD WER, thus, can be analytically computed as $\text{WER}_{\text{MLD}}(p) = \sum_{w=9}^{n} \binom{n}{w} p^w (1-p)^{n-w}$. In this context, therefore, the MLD can be considered as a bounded distance decoder (BDD) with error correcting capacity of $t = 8$. The solid line with $*$ symbols denote the BFD WER computed via the IIS method with $w_{\text{min}} = 1$ and $w_{\text{max}} = 34$. We notice that BFD has a significant E_b/N_0 gain (≥ 0.5 dB) over MLD at a WER of 10^{-10} and ≥ 4 orders of magnitude of WER gain at higher values of E_b/N_0. It is clear from Figure VI.3 that the analytically computable WER_{MLD} does not serve as a good estimate of the BFD WER.

Additionally, in Figure VI.3, the dotted line with o symbols denotes the BFD BER computed via the IIS technique from the same range of Hamming weights. Finally, the squares indicate the BFD WERs obtained via standard MC simulations. The 99% confidence intervals based on the Gaussian assumption [36] were computed for the standard MC WER, although they have not been plotted in figure 2 as they were practically imperceptible on the plotting scale. For all SNR values for which standard MC results were obtained, the corresponding IIS BFD WERs always fell within the corresponding confidence interval.

2. (256,81) 4D SPCP code

Next, consider the $n = 256$, $k = 81$ 4D SPCP code [12]. Decoding of the 4D SPCP code under BFD exhibits different behavior as compared to the EG code decoding discussed previously. Also, as we will see, computation of the

IIS estimator variance for the 4D SPCP code demonstrates the utilization of equation (VI.28) in its entirety; unlike the EG code, the set $\{w', \ldots, w^*\}$ in this case is non-empty. Here, again, we assume BPSK-AWGN transmission followed by binary decisioning.

Table VI.3 lists the word error events and total words transmitted at each w considered. Zero codewords were decoded for all w not in the table. From Table VI.3, it is clear that $w_{\max} = 34$ in this case also. Figure VI.4 shows the raw WER for the different w, and the corresponding weighted WERs at 3, 5, 8, and 11dB.

Here, also, the raw WER is monotonically non-decreasing with w, but becomes close to unity only near $w = 34$. At the smallest SNR of 3dB, the relative bias upper bound computed via equation (VI.19) gives $B_r(\hat{P}(E)) \leq \frac{4.13 \times 10^{-1}}{2.23 \times 10^{-1}} = 1.852$, which is intolerably large. Thus, our IIS results do not provide an accurate estimate of WER at 3dB. This fact also follows from the observation that the weighted WER curve for 3dB in Figure VI.4 barely reaches its peak at $w_{\max} = 34$. As a result, the IIS technique underestimates the WER, as is clearly evident from Figure VI.5 when one compares the standard MC WER estimate with that of IIS.

On the other hand, we observe from Figure VI.4 that the weighted WER curve for 5dB peaks at $w = 25$ and drops to about an order of magnitude below the peak value at $w_{\max} = 34$. Further, the relative bias upper bound at 5dB is computed as $B_r(\hat{P}(E)) \leq \frac{1.015 \times 10^{-3}}{5.18 \times 10^{-2}} = 1.96 \times 10^{-2}$, which is small. Thus, for the 4D SPCP code considered, we conclude that the choice of $w_{\max} = 34$ leads to negligible IIS-estimator bias for SNR $\geq$ 5dB.

The 4D SPCP code being considered has $d_{\min} = 16$ and, hence, the ability to

Table VI.3

Total error patterns transmitted and corresponding word error events recorded at the considered Hamming weights for the $n = 256$, $k = 81$ 4D SPCP code.

Hamming weight, w	1	2	3	4	5	6	7	8	9
N_w	256	$\binom{256}{2}$	$\binom{256}{3}$	$\binom{256}{4}$	10^8	10^8	10^7	10^7	10^7
E_w	0	0	0	0	0	13	14	61	227

Hamming weight, w	10	11	12	13	14	15	16	17	18
N_w	10^7	10^6	10^6	10^6	10^6	10^6	10^6	10^6	10^6
E_w	590	147	298	678	1212	2252	3708	6125	9663

Hamming weight, w	19	20	21	22	23	24	25	26	27
N_w	10^6	10^6	10^6	10^6	10^6	10^6	10^4	10^4	10^4
E_w	14575	21819	31920	45193	62584	85594	1183	1499	1887

Hamming weight, w	28	29	30	31	32	33	34		
N_w	10^4	10^4	10^4	10^4	10^4	10^4	10^4		
E_w	2401	2892	3627	4320	5064	5745	6491		

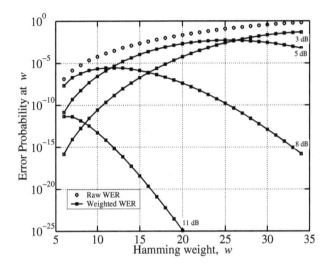

Figure VI.4: Weighted WER at various E_b/N_0 and raw WER vs. Hamming weight w for the $(256, 81)$ 4D SPCP code under BFD

correct ≤ 7 errors via BDD. This code also has an **H**-matrix with 4 ones in each row and column, with the set of PCEs orthogonal on every bit-position. The BFD, thus, is assured of correcting all error-patterns of $w \leq 2$. Additionally, exhaustive error-pattern generation and decoding via BFD reveals that all error-patterns of $w = 3$ and $w = 4$ are successfully decoded to the all-zeros codeword. We observe decoder errors, however, when $w = 6$ and $w = 7$ error-patterns are decoded via BFD (see Table VI.3). Thus, it is clear from the IIS results that the BFD, while able to correct more errors than indicated by the majority-logic decoding bound, is not able to correct all errors up to the error correcting capability of the code.

As mentioned previously, in order to irrefutably determine if error-patterns at a particular weight w result in decoder error or not, one must systematically generate error patterns at that w until either an error event is observed, or until all the error-patterns have been exhausted without any error being observed. Unfortunately, exhaustive error-pattern generation becomes prohibitive as its weight increases. As an example, for this code, we were able to exhaustively generate and decode all $\binom{256}{4} = 174,792,640$ error-patterns of $w = 4$. This exercise took approximately 8.5 days on a 2.4 GHz Intel-Xeon processor. Exhaustive error-pattern evaluation for $w = 5$ would require generating and decoding $\binom{256}{5} = 8,809,549,056$ error-patterns – an impractical task that would require about 430 days on the same machine.

Thus, for this code, the only solution is to randomly generate a large number of error-patterns at each $w \geq 5$ and record the corresponding number of errors. While errors were recorded from randomly generated patterns with $w \geq 6$, evaluation of 10^8 randomly generated $w = 5$ patterns did not result in a single error. Consequently, we find ourselves in a situation where $\hat{P}(E|w=5) = 0$, but we are not certain if $P(E|w=5) = 0$. Thus, based on previously developed notations, we have $w' = w^* = 5$ and the set $\{w', \ldots, w^*\}$ is non-empty.

Equation (VI.28) can be used to bound the IIS estimator variance, for this case as

$$var\{\hat{P}(E)\} \lessapprox \sum_{w=5}^{5} P_w^2 \cdot \frac{\beta}{N_w^2} + \sum_{w=6}^{34} P_w^2 \cdot \frac{\hat{P}(E|w)}{N_w} \ . \tag{VI.32}$$

As indicated previously, we choose $\beta = 5$ to ensure a 99% confidence on the bound in equation (VI.32). The standard-deviation values corresponding to the evaluation of the square-root of equation (VI.32) for some SNRs are given in Table VI.4.

At this point it is worthwhile mentioning that although we have not been able to conclusively prove the ability of BFD to decode all $w = 5$ error-patterns for the considered 4D SPCP code, we have strong reason to believe this to be true. Our belief is based on the following. First, by virtue of the 4-dimensional spatial symmetry of the code, one would expect that if a certain $w = 5$ error pattern fails to decode to the all-zeros codeword, all those $w = 5$ error patterns, which are equivalent to the original pattern due to symmetry will also suffer the same fate. In other words, it is extremely unlikely that we will have exactly one $w = 5$ error pattern that fails to decode to the all-zeros codeword. Second, for each $w = 6$ error-pattern that failed to decode to the all-zeros codeword, we performed BFD decoding on the six distinct $w = 5$ sub-patterns[6], and in every case we observed that the sub-pattern decoded successfully to the all-zeros codeword.

---

[6]A $w = 5$ sub-pattern of $w = 6$ error-pattern is obtained by eliminating the error at one of

If we assume for the moment that the BFD for the considered 4D SPCP code is capable of decoding all $w = 5$ error-patterns to the all-zeros codeword, we can compute the upper bound on the IIS estimator variance by

$$var\{\hat{P}(E)\} \lessapprox \sum_{w=6}^{34} P_w^2 \cdot \frac{\hat{P}(E|w)}{N_w} . \tag{VI.33}$$

Table VI.4 also lists the IIS estimator standard-deviation for this code computed via the square-root of equation (VI.33). We see that the inability to conclusively state that the BFD can decode all w=5 error patterns to the all-zeros codeword affects the IIS estimator standard-deviation markedly at higher SNRs (say 11dB) where $w = 5$ error patterns likely dominate the WER.

In Table VI.4, we also list the standard MC estimator standard-deviation based on the transmission and decoding of $N = \sum_{w=1}^{w_{max}} N_w = 4.317 \times 10^8$ received-words, computed via the square root of the right hand side of equation (VI.31). Here, as previously, we assume that the true probability of error equals that determined by the IIS estimator.

From Table VI.4, we observe that at 3 and 5dB, the standard MC estimator has smaller standard-deviation than the IIS estimator, and all estimator standard-deviations are a few orders of magnitude smaller than the estimated WER. The smaller standard-deviation of the standard MC estimator is primarily due to the large value of transmitted codewords N that we assumed. At 8dB we observe that the IIS estimator standard-deviation (using either equation (VI.32) or equation (VI.33)) is smaller than that of the standard MC estimator. Again,

---

the six different error-positions. Clearly, six different $w = 5$ sub-patterns exist for every $w = 6$ error-pattern.

Table VI.4

The WER estimate, upper bound on relative bias via equation (VI.19), approximate estimator standard-deviation via equation (VI.32) and via equation (VI.33) for IIS for different values of E_b/N_0, and standard-deviation of the MC estimator for $N = 4.317 \times 10^8$ assuming true WER is as determined by the IIS estimator. All results for the $(256, 81)$ 4D SPCP code.

E_b/N_0	3	5	8	11
IIS WER estimate	2.23×10^{-1}	5.18×10^{-2}	1.77×10^{-5}	1.11×10^{-11}
IIS rel. bias upper bound	1.85	1.96×10^{-2}	2.38×10^{-12}	1.39×10^{-38}
IIS estimator std. dev. via (VI.32)	3.16×10^{-4}	4.91×10^{-5}	5.60×10^{-8}	8.31×10^{-12}
IIS estimator std. dev. via (VI.33)	3.16×10^{-4}	4.91×10^{-5}	5.59×10^{-8}	8.42×10^{-15}
Std. MC estimator std. dev. (IIS WER assumed true)	2.27×10^{-5}	1.09×10^{-5}	2.02×10^{-7}	1.60×10^{-10}

all estimator standard-deviations are a few orders of magnitude smaller than the estimated WER.

The situation at 11dB, however, is very different. First, the standard MC estimator standard-deviation is larger than the estimated WER, indicating that standard MC with the specified number of transmitted words is inadequate to estimate this low WER. Second, the IIS estimator standard-deviation based on equation (VI.32) is only slightly smaller than the estimated WER. The use of equation (VI.32) indicates incomplete knowledge about the ability of the BFD to successfully correct $w = 5$ error patterns, and the assumption that the error-rate for $w = 5$ patterns is $\frac{\beta}{N}$. The fact that the IIS estimator standard-deviation via equation (VI.32) is only slightly smaller than the estimated WER indicates that the assumed β errors at $w = 5$ begin to dominate the WER at SNRs ≥ 11

dB. Thus, incomplete knowledge about the decoder performance for $w = 5$ error patterns affects the IIS estimate reliability at SNRs ≥ 11 dB. On the other hand, if one assumes complete knowledge about the ability of the BFD to successfully decode $w = 5$ error patterns (i.e., computation via equation (VI.33)), we see that the standard-deviation is a few orders of magnitude smaller than the WER estimate, indicating high reliability.

Importantly, we observe that in the event of incomplete knowledge of the decoder performance at certain w, it is still possible for the IIS-estimator to provide a wider range of SNR and WER values for which we have reliable estimates, as compared to the range provided by the standard MC estimator with the same total number of transmitted words.

Finally, in this case, the lowest WER that can be estimated by standard MC, using $N = 4.317 \times 10^8$ received-words and adequate reliability, is $\approx \frac{10}{4.317 \times 10^8} = 2.3 \times 10^{-8}$. Estimating lower WERs by standard MC would require many more received-words to be decoded.

Figure VI.5 shows plots of various error rates as a function of E_b/N_0 in dB. The dots represent $p = Q\left(\frac{1}{\sigma}\right)$ for the BSC(p). The dashed line with + symbols shows the BDD WER based on successful decoding of all received-words with $w \leq 7$, with the assumption that no received-word with $w \geq 8$ is decoded successfully. The BDD WER, thus, can be analytically computed as $\text{WER}_{\text{BDD}}(p) = \sum_{w=8}^{n} \binom{n}{w} p^w (1-p)^{n-w}$. The solid line with * symbols denotes the BFD WER computed via the IIS technique with $w_{\min} = 1$ and $w_{\max} = 34$, while the dotted line with ○ symbols denotes the corresponding BER. Finally, the squares indicate the BFD WER obtained via standard MC simulations. Again, the 99% confidence intervals were computed for the standard MC WER, but have not been plotted in Figure VI.5. For all SNRs $\geq$ 5dB for which standard MC results were obtained, the corresponding IIS BFD WERs always fell within the corresponding confidence interval.

The IIS estimates for SNRs < 5dB are not accurate due to the large relative bias. This is also evidenced by the deviation of the IIS estimates from the standard MC estimates at these SNRs. Note that these standard MC estimates can be considered to be very accurate since the WER is very high.

Although the BFD cannot successfully correct all $w = 6$ and $w = 7$ error-patterns, we observe that up to SNRs of about 13 dB, the BFD performs much better than the BDD that can correct all $w \leq 7$ error-patterns. In fact, the BFD has gains of more than 2, 1, and 0.5 dB over the BDD at WERs of 10^{-5}, 10^{-10}, and 10^{-15}, respectively. Such behavior is primarily due to the ability of the BFD to successfully decode a large fraction of error patterns at $8 \leq w \leq 20$. For example, at $w = 8$, the BFD can correct all error-patterns but a fraction that approximately equals 6.1×10^{-6}, and at $w = 20$, the BFD can still correct all error-patterns but a fraction that approximately equals 0.022. Meanwhile, the BDD is assumed to be incapable of correcting even a single error pattern having $w > 7$. As in the case of the EG code, the analytically computable WER_{BDD} does not serve as a good estimate of the BFD WER.

Having discussed the application of IIS to evaluate the performance of codes of higher minimum distance as compared to the RCD codes, we now turn our attention to the evaluation of the $\eta = 17$ and $\eta = 37$ RCD codes. Recall that the analysis of error patterns of RCD codes of specific Hamming weights was instrumental in the conception of the IIS technique.

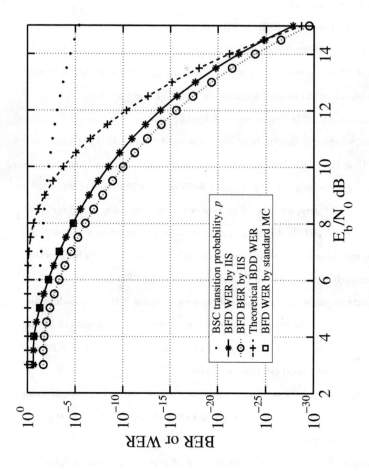

Figure VI.5: Error Probabilities as a function of E_b/N_0 for the $(256, 81)$ 4D SPCP code

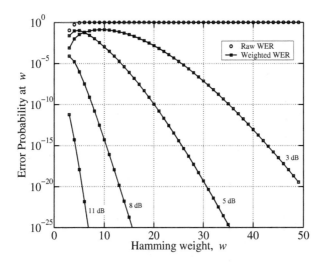

Figure VI.6: Weighted WER at various E_b/N_0 and raw WER vs. Hamming weight w for the $n = 289$, $k = 240$ RCD code with $\eta = 17$ under BFD.

3. $(289, 240)$ $\eta = 17$ **RCD code**

We shall now consider the $\eta = 17$ RCD code with $n = 289$, $k = 240$, and, as previously, we shall assume BPSK-AWGN transmission followed by binary decisioning. Table VI.5 lists the word error events and total words transmitted at each of the Hamming weights considered. Zero words were decoded for all Hamming weights not in the table. The total number of words decoded for the IIS simulation were $N = 4.28 \times 10^7$, and $w_{\max} = 49$. Figure VI.6 shows the raw WER for the different Hamming weights, and the corresponding weighted WERs at 3, 5, 8, and 11dB.

We see that the raw WER is monotonically non-decreasing with w and is very close to unity when $w = 7$. At the smallest SNR of 3dB, the relative

Table VI.5

Total error patterns transmitted and corresponding word error events recorded at the considered Hamming weights for the $n = 289$, $k = 240$ RCD code with $\eta = 17$.

Hamming weight, w	1	2	3	4	5	6	7
N_w	289	$\binom{289}{2}$	$\binom{289}{3}$	10^7	10^7	10^7	10^6
E_w	0	0	430032	5361948	8947763	9879899	999127

Hamming weight, w	8	9	10	11	12	13	14
N_w	10^6	10^6	10^6	10^6	10^6	10^6	10^6
E_w	999964	999997	10^6	10^6	10^6	10^6	10^6

Hamming weight, w	15–19	20–49
N_w	10^5	10^4
E_w	10^5	10^4

Table VI.6

The WER estimate, upper bound on relative bias via equation (VI.19), and approximate estimator standard-deviation via equation (VI.34) for IIS for different values of E_b/N_0, and standard-deviation of the MC estimator for $N = 4.28 \times 10^7$ assuming true WER is as determined by the IIS estimator. All results for the $(289, 240)$ RCD code with $\eta = 17$.

E_b/N_0	3	5	8	11
IIS WER estimate	9.77×10^{-1}	3.19×10^{-1}	9.75×10^{-5}	6.00×10^{-12}
IIS rel. bias upper bound	1.95×10^{-22}	2.4×10^{-45}	3.0×10^{-106}	1.0×10^{-223}
IIS estimator std. dev.	9.89×10^{-5}	2.77×10^{-5}	3.29×10^{-9}	6.81×10^{-20}
Std. MC estimator std. dev.	2.27×10^{-5}	7.12×10^{-5}	1.5×10^{-6}	3.74×10^{-10}
(IIS WER assumed true)				

bias upper-bound computed via equation (VI.19) gives $B_r(\hat{P}(E)) \leq \frac{1.91 \times 10^{-22}}{9.77 \times 10^{-1}} = 1.95 \times 10^{-22}$, which is certainly negligible. Further, the relative bias decreases with increasing SNR (see Table VI.6). Thus, for the SNR values considered, the choice of $w_{\max} = 49$ leads to negligible bias, thereby justifying our choice. In fact, even a choice of $w_{\max} = 20$ would have adequately sufficed for estimating WER at $E_b/N_0 \geq 3$dB, since at $w = 20$ the weighted WER curve for 3dB in Figure VI.6 is decreasing and about two orders of magnitude below the peak value of the curve (the curve peaks at about $w = 10$).

The BFD has been previously shown to be capable of correcting all error patterns up to weight two for the class of RCD codes. The absence of recorded error events during the exhaustive generation and decoding of single and double-error patterns further confirms this fact. Additionally, we have also recorded a significant number of error events during the transmission and decoding of words of Hamming weight $w \geq 3$. As such, the evaluation of the variance of the IIS estimator for this RCD code using equation (VI.28) poses little problem.

Similar to the $(255,175)$ EG code, the first term on the right of equation (VI.28) vanishes in the case of this RCD code (and, in general, for all RCD codes), since for all w for which $\hat{P}(E|w) = 0$, it is known with certainty that $P(E|w) = 0$ (i.e., the set $\{w', \ldots, w^*\}$ is empty). Thus, the estimator variance satisfies

$$var\{\hat{P}(E)\} \lessgtr \sum_{w=3}^{w_{\max}} P_w^2 \cdot \frac{\hat{P}(E|w)}{N_w} . \qquad (VI.34)$$

The standard-deviation computation via equation (VI.34) for various SNR values are given in Table VI.6.

For the E_b/N_0 values considered, the IIS estimator standard-deviation is seen to decrease as E_b/N_0 increases for the same values of N_w and $\hat{P}(E|w)$. Also, relative to the estimated WER value, the IIS estimator standard-deviation is a few orders of magnitude lower for all E_b/N_0 considered, indicating sufficient estimation accuracy.

Next, we compare the IIS estimator standard-deviation with that of a standard MC estimator for the same total number of decoded received-words ($N = \sum_{w=1}^{w_{\max}} N_w == 4.28 \times 10^7$). Assuming, as before, that the true probability of error equals that determined by the IIS estimator, we can compute the standard-deviation of the standard MC estimator for $N = 4.28 \times 10^7$ decoded received-words via the square-root of equation (VI.31). These values are listed in the last row of Table VI.6, where we see that, for $E_b/N_0 \geq$ 5dB, the IIS estimator gives lower standard-deviation as compared to the standard MC estimator, with the gap between the two standard-deviations increasing with increasing E_b/N_0. The smaller standard-deviation of the standard MC estimator at $E_b/N_0 =$ 3dB is primarily due to the large value of transmitted codewords N relative to the WER being estimated. We also note that at 11dB the standard-deviation of the standard MC estimator exceeds the IIS-estimated WER, indicating that the standard MC estimator is ineffective at estimating this low WER. Finally, the lowest WER that can be estimated by standard MC, using $N = 4.28 \times 10^7$ received-words and adequate reliability, is $\approx \frac{10}{4.28 \times 10^7} = 2.34 \times 10^{-7}$. Estimating lower WERs by standard MC would require many more received-words to be decoded.

Figure VI.7 shows plots of various error rates as a function of E_b/N_0 in dB. The dots represent $p = Q\left(\frac{1}{\sigma}\right)$ for BSC(p). The dashed line with $+$ symbols shows the BDD WER based on successful decoding of all received-words of $w \leq 2$, with the assumption that no received-word with $w \geq 3$ is decoded successfully. The BDD WER, thus, can be analytically computed as $\text{WER}_{\text{BDD}}(p) = \sum_{w=3}^{n} \binom{n}{w} p^w (1-p)^{n-w}$. The solid line with $*$ symbols denote the BFD WER computed via the IIS method with $w_{\min} = 1$ and $w_{\max} = 49$. We notice that BFD has a E_b/N_0 gain (≈ 0.3 dB) over BDD at a WER of 10^{-10} and ≈ 1 order of magnitude of WER gain at higher values of E_b/N_0. It is clear from Figure VI.7 that the analytically computable WER_{BDD} does not serve as a good estimate of the BFD WER.

The dotted line with $\circ$ symbols denotes the BFD BER computed via the IIS technique from the same range of Hamming weights, while the solid and dash-dot lines indicate upper and lower bounds on the BFD WER obtained by the theoretical evaluation presented in Chapter III (equations III.37 and III.39, respectively). Finally, the squares indicate the BFD WERs obtained via standard MC simulations. As previously, the 99% confidence intervals were computed for

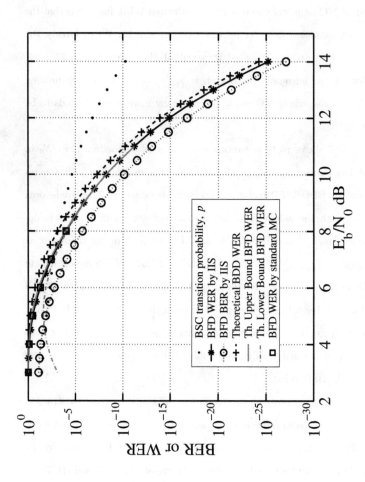

Figure VI.7: Error Probabilities as a function of E_b/N_0 for the $(289, 240)$ RCD code with $\eta = 17$.

the standard MC WER but have not been plotted in Figure VI.7. For all SNR values for which standard MC results were obtained, the corresponding IIS BFD WERs always fell within the corresponding confidence interval.

We also point out that the fraction of received-words at $w = 4$ that lead to error (includes decoder failure and decoding to wrong codewords) for the IIS simulation equals $\frac{5361948}{10^7} = 0.5362$. This fraction is very close to the fraction of received-words at $w = 4$ that lead to error based on an exhaustive enumeration and decoding of all 4-error patterns for the $\eta = 17$ RCD code, which equals $\frac{152574660}{284660376} = 0.5360$ (see Table III.3). As expected, the IIS results agree excellently with the theoretical upper and lower bounds curves that are computed for BFD of the $\eta = 17$ RCD code.

Finally, we also note that the $\eta = 17$ RCD code under BFD achieves a WER of 10^{-10} at $E_b/N_0 \approx 10.5$dB, while under SPA decoding the same WER was achieved at ≈ 8dB (see V.4). This implies a coding gain of about 2.5 for soft-decision decoding over hard-decision decoding.

4. $(1369, 1260)$ $\eta = 37$ RCD code

Finally, we present results for the $\eta = 37$ RCD code with $n = 1369$, $k = 1260$. BPSK-AWGN transmission followed by binary decisioning is assumed, and results are presented along the same lines as those for the $\eta = 17$ RCD code. Table VI.7 lists the word error events and total words transmitted at each of the Hamming weights considered. For all Hamming weights not in the table, no words were decoded. The total number of words decoded for the IIS simulation were $N = 1.38 \times 10^7$, and $w_{max} = 79$. Figure VI.8 shows the raw WER for the different Hamming weights, and the corresponding weighted WERs at 3, 5, 8,

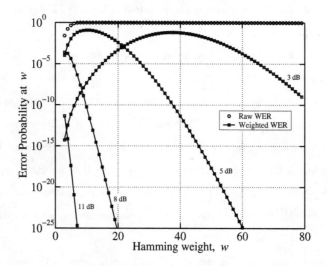

Figure VI.8: Weighted WER at various E_b/N_0 and raw WER vs. Hamming weight w for the $n = 1369$, $k = 1260$ RCD code with $\eta = 37$ under BFD.

and 11dB.

We see that the raw WER is monotonically non-decreasing with w and is very close to unity when $w = 8$. At the smallest SNR of 3dB, the relative bias upper-bound computed via equation (VI.19) gives $B_r(\hat{P}(E)) \leq \frac{1.705 \times 10^{-10}}{1} = 1.705 \times 10^{-10}$, which is certainly negligible. This, coupled with the fact that the relative bias decreases with increasing SNR (see Table VI.8) implies that, for the SNR values considered, the choice of $w_{\max} = 79$ leads to negligible bias. In fact, an examination of Figure VI.8 reveals that even a choice of $w_{\max} = 60$ would have adequately sufficed for estimating WER at $E_b/N_0 \geq 3$dB.

Equation VI.34 can also be used to compute the estimator variance for the $\eta = 37$ RCD code. The standard-deviation computation via equation (VI.34)

Table VI.7

Total error patterns transmitted and corresponding word error events recorded at the considered Hamming weights for the $n = 1369$, $k = 1260$ RCD code with $\eta = 37$.

Hamming weight, w	1	2	3	4	5	6	7
N_w	1369	$\binom{1369}{2}$	10^6	10^6	10^6	10^6	10^6
E_w	0	0	25457	166779	450992	767059	948020

Hamming weight, w	8	9	10	11	12	13	14
N_w	10^6	10^6	10^6	10^6	10^6	10^6	10^6
E_w	994355	999608	999981	999999	10^6	10^6	10^6

Hamming weight, w	15–21	22–29	30–79
N_w	10^5	10^4	10^3
E_w	10^5	10^4	10^3

Table VI.8

The WER estimate, upper bound on relative bias via equation (VI.19), and approximate estimator standard-deviation via equation (VI.34) for IIS for different values of E_b/N_0, and standard-deviation of the MC estimator for 1.38×10^7 assuming true WER is as determined by the IIS estimator. All results for the $(1369, 1260)$ RCD code with $\eta = 37$.

E_b/N_0	3	5	8	11
IIS WER estimate	1	9.58×10^{-1}	4.71×10^{-4}	4.40×10^{-12}
IIS rel. bias upper bound	1.7×10^{-10}	5.65×10^{-44}	6.3×10^{-150}	-
IIS estimator std. dev.	1.57×10^{-3}	1.01×10^{-4}	1.50×10^{-7}	3.62×10^{-19}
Std. MC estimator std. dev.	8.27×10^{-9}	5.41×10^{-5}	5.84×10^{-6}	5.65×10^{-10}
(IIS WER assumed true)				

for various SNR values are given in Table VI.8. Relative to the estimated WER value, the IIS estimator standard-deviation is a few orders of magnitude lower for all E_b/N_0 considered, indicating sufficient estimation accuracy.

Based on decoding $N = \sum_{w=1}^{w_{\max}} N_w == 1.38 \times 10^7$ received-words, the MC estimator standard-deviation can be computed via the square-root of equation (VI.31), assuming, as before, that the true probability of error equals that determined by the IIS estimator. These values are listed in the last row of Table VI.8, where, we see that for the higher E_b/N_0 values (8 and 11dB), the IIS estimator gives lower standard-deviation as compared to the standard MC estimator. On the other hand, the standard-deviation of the MC estimator at the lower E_b/N_0 values (3 and 5dB) is smaller than that of the IIS estimator due to the inordinately large value considered for N in relation to the WER being estimated. Finally, the lowest WER that can be estimated by standard MC, using $N = 1.38 \times 10^7$ received-words and adequate reliability, is $\approx \frac{10}{1.38 \times 10^7} = 7.26 \times 10^{-7}$. Estimating lower WERs by standard MC would require many more received-words to be decoded.

Figure VI.9 shows plots of various error rates as a function of E_b/N_0 in dB. The dots, the solid line with $*$ symbols, the dotted line with $\circ$ symbols, the solid line, the dash-dot line, and the squares, all have their usual meanings. The dashed line with $+$ symbols shows the BDD WER based on successful decoding of all received-words of $w \leq 2$, with the assumption that no received-word with $w \geq 3$ is decoded successfully. The BDD WER, thus, can be analytically computed as $\text{WER}_{\text{BDD}}(p) = \sum_{w=3}^{n} \binom{n}{w} p^w (1-p)^{n-w}$. We notice that BFD has a E_b/N_0 gain (≈ 0.4 dB) over BDD at a WER of 10^{-10} and ≈ 2 orders of magnitude of WER gain at higher values of E_b/N_0. It is clear from Figure VI.9 that the analytically computable WER_{BDD} does not serve as a good estimate of the BFD WER.

As previously, the 99% confidence intervals were computed for the standard MC WER, but have not been plotted in Figure VI.9. For all SNR values for which standard MC results were obtained, the corresponding IIS BFD WERs always fell within the corresponding confidence interval. The IIS results also agree excellently with the theoretical upper and lower bounds curves that are computed for bit-flipping decoding of the $\eta = 37$ RCD code.

We also note that the $\eta = 37$ RCD code under BFD achieves a WER of 10^{-10} at $E_b/N_0 \approx 10.5$dB, while under SPA decoding the same WER was achieved at ≈ 8dB (see Figure V.4). This implies a coding gain of about 2.5 for soft-decision decoding over hard-decision decoding for this code.

Strikingly, all the numbers presented in this discussion for the $\eta = 37$ are the same as the corresponding numbers for the $\eta = 17$ code. In fact, we observe

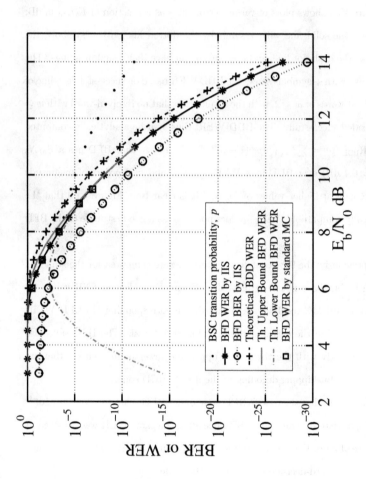

Figure VI.9: Error Probabilities as a function of E_b/N_0 for the (1369, 1260) RCD code with $\eta = 37$.

that the performance of different RCD codes is approximately the same over a range of E_b/N_0 for both SPA decoding (6 to 10 dB) as well as for BFA decoding (9 to 12 dB).

From the IIS results for the two RCD codes, we see that the WER performance curves for BFD and BDD at very high SNRs have fixed vertical separation. This follows since both the curves are dominated by the probability of errors due to 3-error patterns at high SNR. The separation between the two curves occurs due to the large number of 3-error patterns that can be decoded successfully by the BFA; this number is assumed to be 0 for the BDD. In fact, it is easily verified that for high SNR the vertical separation between the two curves corresponds to a factor that equals the raw WER for Hamming weight 3. The raw WER at Hamming weight 3, luckily, decreases with increasing η, and thus longer RCD codes will have larger vertical separation between the WER curves corresponding to decoding via BFD and BDD.

The examples presented in this section demonstrate the utility of the IIS technique in accurately estimating WER and BER of codes of moderate length. We emphasize that, if we have complete knowledge of the ability of the decoder to correct all error patterns with Hamming weights for which the IIS estimator yields zero errors, then we can in principle accurately estimate the BFD WER down to any arbitrarily low value based on the computed raw WERs.

F. Conclusions

1. Summary and conclusions

We presented the concept of a hard-decision FEC decoder output's SNR-invariance to the binary symmetric channel (BSC) transition probability p (defined by the SNR) given a received length-n binary word. We then used this concept of SNR-invariance to develop an importance sampling (IS) technique (denoted invariance importance sampling (IIS)) for hard-decision decoding performance evaluation of binary linear block codes. The equations describing IIS were presented and examples demonstrating the IIS technique were provided. An analysis of the IIS estimator accuracy was provided in terms of its relative bias and variance. We then presented results of IIS simulations performed for the $n = 255$, $k = 175$ Euclidean geometry FEC code, the $n = 256$, $k = 81$ 4-dimensional single-parity-check product FEC code, and the $\eta = 17$ and $\eta = 37$ RCD codes under bit-flipping decoding (BFD). We also evaluated the accuracy of the estimators for each code.

The IIS technique finds its roots in the analysis of 3- and 4-error patterns for the RCD codes that was presented in Chapter III. Based on the examples presented, we conclude that it is possible to use the proposed IIS technique to accurately (with negligible relative bias and small variance) evaluate word-error rate (WER) and bit-error rate (BER) to arbitrarily low values for the chosen codes. In fact, a significant advantage of the IIS technique is that under appropriate conditions, such as small-to-moderate code lengths or when knowledge of the error correcting capability of the decoder is known, the results obtained from a single IIS simulation are sufficient for accurately evaluating WERs and BERs for a range of SNRs, where the upper end of the range may often be any arbitrarily chosen high SNR value. We also conclude that even without definitively knowing whether the decoder can correct all error-patterns at certain Hamming weights, it is still possible for the IIS-estimator to improve the range of SNR and WER for which we have reliable estimates, as compared to the standard MC estimator with the same total number of transmitted words. Further, in such a case, the probability estimates obtained via IIS always serve as a lower bound on the true probability values.

The IIS simulation results presented in this chapter were all obtained in reasonable time frames on commonly available current technology workstations (e.g., a dedicated Pentium-4 2.4 GHz 2GB RAM machine). The total number of decoded words and the approximate times taken for the various codes are provided in Table VI.9. This data should be used only as a rough guideline. Actual times will vary from computer to computer. The number of words that must be decoded and, hence, the time required for standard Monte Carlo simulations is inversely proportional to the WER that must be estimated; the time requirements for standard Monte Carlo simulations for estimating low WERs of the order of 10^{-12} approximately runs into hundreds of years on similar computers. For lower WERs, the time requirements of standard Monte Carlo are only higher.

For a particular RCD code, the vertical separation between the BFD WER curve and the bounded-distance decoder (BDD) WER curve at a specific high E_b/N_0 corresponds to a factor that equals the raw WER at Hamming weight 3 for that code. Since, the raw WER at Hamming weight 3 decreases for increasing η for the class of RCD codes, we conclude that at a high value of E_b/N_0 the WER

Table VI.9

The total number of decoded error-patterns and time taken for IIS simulations for the various codes presented in this chapter.

Code (n, k)	Error-patterns decoded for IIS	Time taken (hours)
$(255, 175)$ EG	2.33×10^6	8.5
$(256, 81)$ 4-D SPCP	4.31×10^8	580
$(289, 240)$ $\eta = 17$ RCD	4.28×10^7	76
$(1369, 1260)$ $\eta = 37$ RCD	1.38×10^7	170

improvement obtained by BFD decoding over BDD decoding for an RCD code is larger for larger η. In other words, RCD codes with larger η derive more benefit at high E_b/N_0 from using the BFD instead of the BDD.

2. Future research

Future work in this area involves the following:

1. It would certainly be an interesting exercise to employ the IIS technique to evaluate longer codes with larger minimum distances and under different hard-decision decoding algorithms. This exercise would provide us with a better understanding of the practical extent of applicability of the IIS technique.

2. The decoding analysis of lower-weight sub-patterns of an error-pattern that results in decoder error is another possible area for investigation, as it will lead to more information on the decoder's ability to correct error-patterns at the lower weight.

3. The IIS technique also may be used to evaluate the performance of codes and decoders on the binary symmetric channel with erasures (BSC/E). Such an analysis would allow us to estimate the code performance for a range of BSC/E transition probability values, without resorting to independent simulations for each choice of transition probabilities.

Chapter VII.

QUANTUM CODES FROM CODES

BASED ON DIAGONALS ON

SQUARE ARRAYS

In this chapter we study the application of the FEC code construction technique
discussed in this dissertation to the construction of *quantum error correcting
codes* or *quantum codes*. We first provide a brief introduction to the concept of
quantum error correction (QEC). Then, we outline the Calderbank-Shor-Steane
technique to construct quantum codes. Next, we discuss how the class of codes
based on diagonals in square arrays with prime side length can be used to construct quantum codes via the CSS techniques, and present some heuristic design
rules that are likely to ensure the highest possible dimensionality and minimum
distance for quantum codes constructed from diagonals on square arrays with
prime side length. We also discuss a related technique for the construction of CSS
quantum codes from *weakly dual* classical codes based on even-sloped diagonals
on square arrays of even side length.

We end the chapter with conclusions and areas for future research.

A. Introduction to Quantum Error Correction

The field of *quantum computation, information processing, and communications* has seen tremendous activity in the recent past [59,61]. This field focusses on utilizing the unique properties of systems that exhibit distinct *quantum mechanical* behavior [16] to communicate and process information.

Quantum systems that must be manipulated for the purposes of *quantum computation, information processing, and communications* are *intrinsically delicate* [29] and highly susceptible to perturbations due to interactions with the surroundings. Maintaining a quantum system in a particular state, or having the ability to recover a quantum state from its perturbed (corrupted) version, is absolutely essential if one must ensure the success of the various quantum tasks one intends to perform. Quantum error correction (QEC) refers to the study of the theory and techniques that enable us to overcome the potentially debilitating effects of the disturbances that incessantly attack the delicate balance of quantum states.

The theory of QEC may be considered as the quantum counterpart of the theory of *error correcting codes* [75] that addresses the issue of correcting for errors that occur in classical communications and storage systems. As one may expect, QEC borrows heavily from concepts in classical error correction. There are distinct differences that separate QEC from its classical analog, however, thanks to the intricacies of quantum mechanics and the absence of classical analogs to a plethora of quantum phenomena.

Extensive research carried out mostly in the last 10 years has put the field of QEC on a strong theoretical footing [59, 61]. The QEC problem has been precisely defined and a set of necessary and sufficient conditions for QEC to be possible have been derived; all this, in spite of the seemingly unsurmountable challenges imposed by the laws of quantum mechanics.

[1]QEC codes or quantum codes map an arbitrary quantum state $|\psi\rangle \in \mathcal{H}_\psi$ to an element $|\theta\rangle$ of a same-dimensional subspace C of a larger space $\mathcal{H}_\theta$. Here, $\mathcal{H}_\psi$ and $\mathcal{H}_\theta$ are complex Hilbert spaces [45] of appropriate dimensions. Thus, $|\theta\rangle \in C \subset \mathcal{H}_\theta$ and $\dim(C) = \dim(\mathcal{H}_\psi) < \dim(\mathcal{H}_\theta)$. The subspace C of $\mathcal{H}_\theta$ is called a quantum code.

The perturbations that affect the delicate quantum states are called *quantum noise processes* and may arise due to the process of physically moving the quantum state (e.g., a polarized photon with the quantum state described by a superposition of two orthogonal polarization states passing through an optical fiber) or due to the simple act of storing a quantum state over a period of time (e.g., an electron with the quantum state described by a superposition of two orthogonal spin states interacting with neighboring charged particles over time). In this context, any unitary evolution[2]of the quantum state or interaction with the environment may be regarded as a source of noise. The noise process may be mathematically modeled as a *quantum operation* $\mathcal{E}(\sigma) = \{E_i\}$, where $\sum_k E_k^\dagger E_k \leq I$, and where $\sigma = |\psi\rangle\langle\psi|$ is the *density matrix* corresponding to the quantum state $|\psi\rangle$ (see Chapters 2 and 8 of [59]).

One of the most important theorems in the theory of quantum error correction states that if the noise process described by $\mathcal{E}(\cdot) = \{E_i\}$ can be corrected

---

[1]In the following discussion, we use the standard Dirac notation to represent quantum states. The reader is referred to [59] for a comprehensive discussion on the standard Dirac notation.

[2]Evolution of *closed* quantum systems is described via unitary transformations. See Chapter 2 of [59].

by the quantum code C, then any noise process built from a linear combination of $\{E_i\}$ can also be corrected by C. For many common noise processes, which includes those defined by the *bit-flip* channel, *phase-flip* channel, and the *depolarization* (or *decoherence* [28]) channel, the operator elements $\{E_i\}$ may be written as a linear combination of the Pauli matrices (or Pauli tensors). Thus, in order to be assured of error-correction against such noise processes, it is sufficient to show error-correction capability for these Pauli tensors (see Chapter 8 and Chapter 10 of [59] for details).

We note that not all noisy channels may be interpreted as a random linear combination of the Pauli matrices. An example of such a case is the *amplitude damping* channel [59]. Despite this inadequacy of the theory based on characterizing correctable errors via Pauli tensors, however, QEC codes designed on the basis of Pauli errors can improve the *fidelity*[3] of the communication/storage process on channels similar to the amplitude damping channel.

The theory of QEC provides us with extremely powerful tools that can help us determine if a given QEC code can correct specific quantum errors. The theory does not provide us, however, with any explicit recipe for the construction of QEC codes that may meet certain desired error-correction capability requirements. Consequently, significant research has been performed also in the area of constructing QEC codes.

One of the popular techniques to construct quantum codes is called the

---

[3]Let $|\psi\rangle$ be the input state to the communication/storage process and let ρ be the output state (as a density matrix). The fidelity (a proximity measure) of the process is computed as $f = \sqrt{\langle\psi|\rho|\psi\rangle}$. $0 \leq f \leq 1$ and larger the fidelity, the closer the output to the input(see Chapters 9 and 10 of [59]).

Calderbank-Shor-Steane (CSS) construction technique named after its discoverers. It turns out that the CSS technique to construct quantum codes is a special case of the construction technique based on the concept of *stabilizers* [59].

In the next section we outline the CSS construction technique.

B. The CSS Construction Technique

CSS codes are constructed from a pair of conventional error-control codes that meet certain requirements. Consider a binary linear block code C_1 with length n and number of information bits k,[4] and a second code C_2 with parameters (n, k_2) such that $C_2 \subset C_1$ (i.e., C_2 is a vector subspace of C_1). Further C_1 and $C_2^\perp$ are chosen such that each is capable of correcting t errors[5].

For each $x \in C_1$, we can construct the quantum state

$$|x + C_2\rangle = \frac{1}{\sqrt{|C_2|}} \sum_{y \in C_2} |x + y\rangle , \qquad (\text{VII.1})$$

where the $+$ symbol represents element by element modulo-2 addition, while the $\sum$ represents a superposition of quantum states. The CSS code $\text{CSS}(C_1, C_2)$ is the $2^{k_1 - k_2}$ dimensional vector space (and a subspace of $\mathcal{H}_{2^n}$) spanned by the states $|x + C_2\rangle$, $\forall x \in C_1$. Further, $\text{CSS}(C_1, C_2)$ is a $[n, k_1 - k_2]$ quantum code[6] that is capable of correcting up to t arbitrary qubit errors.

---

[4]Classical codes with length n and k information bits will be referred to as (n, k) codes. If the minimum distance is known to be d, then we use the notation (n, k, d) to denote the code parameters.

[5]The $\perp$ symbol indicates the dual (or orthogonal complement) of a code.

[6]Square brackets are used to designate quantum codes. An $[n, k]$ quantum code encodes k

Since $C_2 \subset C_1$, all $x \in C_1$ that also satisfy $x \in C_2$ will map to the same codeword of $\text{CSS}(C_1, C_2)$ due to the linearity of the code. Similarly, all $x \in C_1$ that belong to the same coset of C_2 will map to the same codeword. Hence, the number of distinct codewords in $\text{CSS}(C_1, C_2)$ is equal to $\frac{|C_1|}{|C_2|}$. A popular example of a CSS code is the $[7,1]$ code constructed with the $(7,4,3)$ Hamming code as C_1 and the $(7,3,4)$ simplex code, the dual of C_1, as C_2. This $[7,1]$ code is capable of correcting any arbitrary single qubit error.

A special case of the CSS technique occurs when $C_1 = C$, $C_2 = C^\perp$, and $C^\perp \subset C$, i.e., when C is a *weakly-dual* (n,k) code.[7] Since $C^\perp$ is an $(n, n-k)$ code, we immediately see that a *weakly-dual* code must have $k \geq \frac{n}{2}$. If $\mathbf{H}$ denotes the parity-check matrix of the binary code C, then the condition for *weak duality* may be equivalently represented as $\mathbf{HH}^T = \mathbf{0} \mod 2$. In other words, $\mathbf{H}$ represents a *weakly-dual* binary code if the two following conditions are satisfied [49]:

(a) each row of $\mathbf{H}$ must have even weight, and

(b) any two rows must have even overlap[8].

Condition (a) ensures that the rows are self orthogonal, and condition (b) ensures that any two different rows are orthogonal to each other.

If the code C with minimum distance d and error correcting capacity $t = \lfloor \frac{d-1}{2} \rfloor$ is *weakly-dual*, then the quantum code $\text{CSS}(C, C^\perp)$ is a $[n, 2k-n]$ quantum

---

qubits into an n qubit codeword.

[7]A classical $(n.k)$ code C is said to be *weakly-dual* (or *dual-containing*) [49] if $C^\perp \subset C$.

[8]Two binary vectors v_1 and v_2 of the same length are said to have even overlap if the number of positions in which both vectors have a 1 is even.

code that can correct up to t arbitrary qubit errors. Thus in order to have a high-rate quantum code based on the CSS technique and *weakly-dual* codes, one must be able to construct high-rate classical codes with large minimum distance and that are also *weakly-dual*. In the construction of classical codes, the conditions that the code must satisfy for high rate directly conflict with the conditions for large minimum distance. The problem of designing codes is further complicated by the *weakly-dual* requirement.

CSS codes have a desirable property that error-detection and correction is achievable by the application of only Hadamard and C-NOT quantum gates [59]. Further, the number of gates required is linear in the length of the code.

In the next two sections we discuss two different methods to construct classical codes from diagonals on square arrays that may be used to construct CSS quantum codes.

C. CSS Quantum Codes from Codes Based on Diagonals in Square Arrays with η Prime

The class of codes based on diagonals in $\eta \times \eta$ square arrays with η prime, which are described in detail in this dissertation (see Chapter II), are naturally suited for the construction of CSS quantum codes, thanks to their inherent structure.

For a specific value of prime η, let $C_s^{(\eta)}$ represent the classical code constructed on the $\eta \times \eta$ array with s different diagonal bundles forming the parity-check equations and $0 \leq s \leq \eta+1$ (recall that there are $\eta+1$ different diagonal bundles, and see Chapter II for a detailed discussion on the construction of codes based on diagonals in square arrays). Using this notation, the RCD code with parameter

η would correspond to $C_3^{(\eta)}$. In general, $C_s^{(\eta)}$ is a $(\eta^2, (\eta-1)(\eta-s+1))$ code.

Let $C_s^{(\eta)\perp}$ represent the dual (orthogonal complement) of $C_s^{(\eta)}$; then $C_s^{(\eta)\perp}$ is a $(\eta^2, \eta^2 - (\eta-1)(\eta-s+1))$ code. Further, let $\mathbf{H}_s^{(\eta)}$ represent the parity-check matrix of $C_s^{(\eta)}$ (thus the generator matrix of $C_s^{(\eta)\perp}$). Every row of $\mathbf{H}_s^{(\eta)}$ has η ones. Thus, $C_s^{(\eta)\perp}$ always has codewords of weight η and, consequently, its minimum distance is $\leq \eta$.

Let us further impose the restriction that the slopes of the s different diagonal bundles to construct $\mathbf{H}_s^{(\eta)}$ are always chosen from the set $\{0, 1, \ldots, \eta-1, \infty\}$ in a specific sequence. That is, we may require that one always first choose the row diagonal bundle (slope-0), followed by the column diagonal bundle (slope-∞), followed by the slope-1 diagonal bundle, followed by the slope-2 diagonal bundle, and so on, until s different diagonal bundles have been chosen. Based on this sequence restriction, we clearly see that for a specific η, $C_i^{(\eta)\perp} \subset C_j^{(\eta)\perp}$ if $j > i$, $0 \leq i, j \leq (\eta+1)$, since $\mathbf{H}_i^{(\eta)}$ will be a sub-matrix of $\mathbf{H}_j^{(\eta)}$. This, in turn, implies that $C_j^{(\eta)} \subset C_i^{(\eta)}$. The preceding fact can be understood easily if we realize that every codeword of $C_j^{(\eta)}$ always satisfies all the constraints that must be satisfied by codewords of $C_i^{(\eta)}$, and some additional constraints.

It is immediately clear from the preceding discussion and the CSS construction technique that the two codes $C_i^{(\eta)}$ and $C_j^{(\eta)}$ for any $j > i$, $0 \leq i, j \leq (\eta+1)$ can be used to construct the quantum code $\text{CSS}\left(C_i^{(\eta)}, C_j^{(\eta)}\right)$. In general, $\text{CSS}\left(C_i^{(\eta)}, C_j^{(\eta)}\right)$ will be an $[\eta^2, (\eta-1)(j-i)]$ quantum code. The error-correcting capability of $\text{CSS}\left(C_i^{(\eta)}, C_j^{(\eta)}\right)$ will be determined by the error-correcting capabilities of $C_i^{(\eta)}$ and $C_j^{(\eta)\perp}$. Similarly, one can also construct $[\eta^2, (\eta-1)(j-i)]$ quantum codes $\text{CSS}\left(C_j^{(\eta)\perp}, C_i^{(\eta)\perp}\right)$ for any $j > i$, $0 \leq i, j \leq (\eta+1)$, with error correcting capability determined again by the error-correcting capabilities of $C_j^{(\eta)\perp}$ and $\left(C_i^{(\eta)\perp}\right)^\perp = C_i^{(\eta)}$.

Let $d_i^{(\eta)}$ and $d_j^{(\eta)\perp}$ be the minimum distances of $C_i^{(\eta)}$ and $C_j^{(\eta)\perp}$, respectively, then the minimum distance of $\mathrm{CSS}\left(C_i^{(\eta)}, C_j^{(\eta)}\right)$ will be

$$d_{\eta,i,j} = \min\left(d_i^{(\eta)}, d_j^{(\eta)\perp}\right),\qquad\text{(VII.2)}$$

and that of $\mathrm{CSS}\left(C_j^{(\eta)\perp}, C_i^{(\eta)\perp}\right)$ will be

$$d_{\eta,i,j}^\perp = d_{\eta,i,j} = \min\left(d_i^{(\eta)}, d_j^{(\eta)\perp}\right).\qquad\text{(VII.3)}$$

Of course, $d_{\eta,i,j} \leq \eta$ since $d_j^{(\eta)\perp} \leq \eta$. For a specific η, a large value of $d_{\eta,i,j}$ may be obtained by a judicious choice of i and j.

We note that a similar concept of using the class of Reed-Muller codes to construct CSS quantum codes has been discussed in Chapter 7 of [61].

Next we provide some examples of such CSS codes for $\eta = 5$ and $\eta = 7$, and discuss judicious choices of i and j. The examples provide us with enough information to lay down certain heuristics to choose i and j.

1.   CSS codes for $\eta = 5$

For $\eta = 5$ it is fairly easy to compute the minimum distances of all the codes $C_s^{(\eta)}$ and their duals $C_s^{(\eta)\perp}$ via an exhaustive generation and enumeration of codewords. We do not consider $s = 0$ and $s = \eta + 1$ since both these situations correspond to trivial codes. Table VII.1 lists various classical codes possible, as well as their duals, based on diagonals on a 5×5 array. Their parameters are listed in the (n, k, d) format, where n is the code length, k is the dimensionality,

Table VII.1

Classical codes based on diagonals in a square array with $\eta = 5$, their duals, and parameters

Code	(n,k,d)	Dual code	(n,k,d)
$C_1^{(5)}$	$(25,20,2)$	$C_1^{(5)\perp}$	$(25,5,5)$
$C_2^{(5)}$	$(25,16,4)$	$C_2^{(5)\perp}$	$(25,9,5)$
$C_3^{(5)}$	$(25,12,6)$	$C_3^{(5)\perp}$	$(25,13,5)$
$C_4^{(5)}$	$(25,8,8)$	$C_4^{(5)\perp}$	$(25,17,4)$
$C_5^{(5)}$	$(25,4,10)$	$C_5^{(5)\perp}$	$(25,21,2)$

and d is the minimum distance of the code. Recall, we need to achieve at least single-error correction capability.

For the classical codes listed in Table VII.1, we have

$$C_5^{(5)} \subset C_4^{(5)} \subset C_3^{(5)} \subset C_2^{(5)} \subset C_1^{(5)}, \tag{VII.4}$$

and we also have

$$C_1^{(5)\perp} \subset C_2^{(5)\perp} \subset C_3^{(5)\perp} \subset C_4^{(5)\perp} \subset C_5^{(5)\perp}. \tag{VII.5}$$

From equations (VII.4) and (VII.5), we can construct a variety of different CSS codes. As an example, since $C_4^{(5)} \subset C_2^{(5)}$, we can construct the code $\text{CSS}\left(C_2^{(5)}, C_4^{(5)}\right)$, which will have length = 25, dimensionality = $16 - 8 = 8$, and minimum distance $d_{5,2,4} = \min\left(d_2^{(5)} = 4, d_4^{(5)\perp} = 4\right) = 4$. Thus $\text{CSS}\left(C_2^{(5)}, C_4^{(5)}\right)$

Table VII.2

Quantum CSS codes with at least single error correcting capability ($d \geq 3$) derived from the codes in Table VII.1.

CSS Code	$[n, k, d]$	CSS Code	$[n, k, d]$
$\text{CSS}\left(C_2^{(5)}, C_3^{(5)}\right)$	$[25, 4, 4]$	$\text{CSS}\left(C_4^{(5)\perp}, C_3^{(5)\perp}\right)$	$[25, 4, 4]$
$\text{CSS}\left(C_2^{(5)}, C_4^{(5)}\right)$	$[25, 8, 4]$	$\text{CSS}\left(C_4^{(5)\perp}, C_2^{(5)\perp}\right)$	$[25, 8, 4]$
$\text{CSS}\left(C_3^{(5)}, C_4^{(5)}\right)$	$[25, 4, 4]$	$\text{CSS}\left(C_3^{(5)\perp}, C_3^{(5)\perp}\right)$	$[25, 4, 4]$

is a $[25, 8, 4]$ quantum code.

Table VII.2 lists the CSS codes, and their $[n, k, d]$ parameters, that are derived from the codes listed in Table VII.1 and have at least single-error correcting capability. Since we are considering quantum codes with at least single error correcting capability ($d \geq 3$), $C_1^{(5)}$ and $C_5^{(5)\perp}$ are not used to construct the quantum codes.

From Table VII.2 it is obvious that the two quantum codes $\text{CSS}\left(C_2^{(5)}, C_4^{(5)}\right)$ and $\text{CSS}\left(C_4^{(5)\perp}, C_2^{(5)\perp}\right)$ have the best parameters in the set, given that all the listed quantum codes have a minimum distance of 4. If we designate the quantum code as $\text{CSS}(C_1, C_2)$, then $\text{CSS}\left(C_2^{(5)}, C_4^{(5)}\right)$ and $\text{CSS}\left(C_4^{(5)\perp}, C_2^{(5)\perp}\right)$ are observed to be the most balanced since C_1 and $C_2^\perp$ (i.e., $C_2^{(5)}$ and $C_4^{(5)\perp}$, and $C_4^{(5)\perp}$ and $C_2^{(5)\perp}$, respectively) have exactly the same minimum distance of 4.

We next present the parameters for some $\text{CSS}\left(C_i^{(7)}, C_j^{(7)}\right)$ codes, and then discuss certain rules-of-thumb that are likely to result in CSS codes with the best rate for a specified minimum distance.

Table VII.3

Classical codes, and their duals and parameters, based on diagonals in a square array with $\eta = 7$.

Code	(n, k, d)	Dual code	(n, k, d)
$C_1^{(7)}$	$(49, 42, 2)$	$C_1^{(7)\perp}$	$(49, 7, 7)$
$C_2^{(7)}$	$(49, 36, 4)$	$C_2^{(7)\perp}$	$(49, 13, 7)$
$C_3^{(7)}$	$(49, 30, 6)$	$C_3^{(7)\perp}$	$(49, 19, 7)$
$C_4^{(7)}$	$(49, 24, 8)$	$C_4^{(7)\perp}$	$(49, 25, 7)$
$C_5^{(7)}$	$(49, 18, 12)$	$C_5^{(7)\perp}$	$(49, 31, 6)$
$C_6^{(7)}$	$(49, 12, 12)$	$C_6^{(7)\perp}$	$(49, 37, 4)$
$C_7^{(7)}$	$(49, 6, 14)$	$C_7^{(7)\perp}$	$(49, 43, 2)$

2. CSS codes for $\eta = 7$

For $\eta = 7$, it also is fairly easy to determine the minimum distances of all the codes $C_s^{(\eta)}$ and their duals $C_s^{(\eta)\perp}$. Here again, we do not consider $s = 0$ and $s = \eta + 1$. Table VII.3 lists various possible classical codes, and their duals and parameters, based on diagonals on a 7×7 array.

For the codes listed in Table VII.3, we have

$$C_7^{(7)} \subset C_6^{(7)} \subset C_5^{(7)} \subset C_4^{(7)} \subset C_3^{(7)} \subset C_2^{(7)} \subset C_1^{(7)} , \qquad \text{(VII.6)}$$

and

$$C_1^{(7)\perp} \subset C_2^{(7)\perp} \subset C_3^{(7)\perp} \subset C_4^{(7)\perp} \subset C_5^{(7)\perp} \subset C_6^{(7)\perp} \subset C_7^{(7)\perp} . \qquad \text{(VII.7)}$$

Table VII.4

Quantum CSS codes with at least single error correcting capability ($d \geq 3$) derived from the codes in Table VII.3.

CSS Code	$[n, k, d]$	CSS Code	$[n, k, d]$
$\mathrm{CSS}\left(C_2^{(7)}, C_3^{(7)}\right)$	$[49, 6, 4]$	$\mathrm{CSS}\left(C_6^{(7)\perp}, C_5^{(7)\perp}\right)$	$[49, 6, 4]$
$\mathrm{CSS}\left(C_2^{(7)}, C_4^{(7)}\right)$	$[49, 12, 4]$	$\mathrm{CSS}\left(C_6^{(7)\perp}, C_4^{(7)\perp}\right)$	$[49, 12, 4]$
$\mathrm{CSS}\left(C_2^{(7)}, C_5^{(7)}\right)$	$[49, 18, 4]$	$\mathrm{CSS}\left(C_6^{(7)\perp}, C_3^{(7)\perp}\right)$	$[49, 18, 4]$
$\mathrm{CSS}\left(C_2^{(7)}, C_6^{(7)}\right)$	$[49, 24, 4]$	$\mathrm{CSS}\left(C_6^{(7)\perp}, C_2^{(7)\perp}\right)$	$[49, 24, 4]$
$\mathrm{CSS}\left(C_3^{(7)}, C_4^{(7)}\right)$	$[49, 6, 6]$	$\mathrm{CSS}\left(C_5^{(7)\perp}, C_4^{(7)\perp}\right)$	$[49, 6, 6]$
$\mathrm{CSS}\left(C_3^{(7)}, C_5^{(7)}\right)$	$[49, 12, 6]$	$\mathrm{CSS}\left(C_5^{(7)\perp}, C_3^{(7)\perp}\right)$	$[49, 12, 6]$
$\mathrm{CSS}\left(C_3^{(7)}, C_6^{(7)}\right)$	$[49, 18, 4]$	$\mathrm{CSS}\left(C_5^{(7)\perp}, C_2^{(7)\perp}\right)$	$[49, 18, 4]$
$\mathrm{CSS}\left(C_4^{(7)}, C_5^{(7)}\right)$	$[49, 6, 6]$	$\mathrm{CSS}\left(C_4^{(7)\perp}, C_3^{(7)\perp}\right)$	$[49, 6, 6]$
$\mathrm{CSS}\left(C_4^{(7)}, C_6^{(7)}\right)$	$[49, 12, 4]$	$\mathrm{CSS}\left(C_4^{(7)\perp}, C_2^{(7)\perp}\right)$	$[49, 12, 4]$
$\mathrm{CSS}\left(C_5^{(7)}, C_6^{(7)}\right)$	$[49, 6, 4]$	$\mathrm{CSS}\left(C_3^{(7)\perp}, C_2^{(7)\perp}\right)$	$[49, 6, 4]$

From equations (VII.6) and (VII.7), we can construct a variety of different CSS codes. Table VII.4 lists the CSS codes, and their $[n, k, d]$ parameters, that are derived from the codes listed in Table VII.3 and have at least single-error correcting capability. Since we are considering quantum codes with at least single error correcting capability ($d \geq 3$), $C_1^{(7)}$ and $C_7^{(7)\perp}$ are not used to construct the quantum codes.

From Table VII.4 we observe that for a minimum distance of 4, the codes $\mathrm{CSS}\left(C_2^{(7)}, C_6^{(7)}\right)$ and $\mathrm{CSS}\left(C_6^{(7)\perp}, C_2^{(7)\perp}\right)$ have the best parameters in the set (i.e., highest dimensionality k), while for a minimum distance of 6, the codes

$\text{CSS}\left(C_3^{(7)}, C_5^{(7)}\right)$ and $\text{CSS}\left(C_5^{(7)\perp}, C_3^{(7)\perp}\right)$ have the highest dimensionality.

If we designate the quantum code as $\text{CSS}(C_1, C_2)$, then we also see that these four codes just listed are the most balanced since C_1 and $C_2^{\perp}$ have exactly the same minimum distance for each quantum code (i.e., $C_2^{(7)}$ and $C_6^{(7)\perp}$, and $C_6^{(7)\perp}$ and $C_2^{(7)}$, respectively have minimum distance of 4, while $C_3^{(7)}$ and $C_5^{(7)\perp}$, and $C_5^{(7)\perp}$ and $C_3^{(7)}$, respectively, have minimum distance of 6).

In the next subsection, we discuss some heuristic design rules based on the observations presented in this and the previous subsections.

3. Design rules for choosing i and j

For both $\eta = 5$ and $\eta = 7$, we see that the quantum codes $\text{CSS}\left(C_i^{(\eta)}, C_j^{(\eta)}\right)$ or $\text{CSS}\left(C_j^{(\eta)\perp}, C_i^{(\eta)\perp}\right)$ that achieved the highest dimensionality (k) for a given minimum distance satisfy the code parameters relationship

$$i + j = \eta + 1 \tag{VII.8}$$

(for $d_{\eta,i,j} = d_{\eta,j,i}^{\perp} = 4$, see $\text{CSS}\left(C_2^{(5)}, C_4^{(5)}\right)$ and $\text{CSS}\left(C_4^{(5)\perp}, C_2^{(5)\perp}\right)$, and $\text{CSS}\left(C_2^{(7)}, C_6^{(7)}\right)$ and $\text{CSS}\left(C_6^{(7)\perp}, C_2^{(7)\perp}\right)$).

We see that for the codes $C_i^{(\eta)}$, the minimum distance roughly increases by 2 each time i is incremented by 1. In each case we also see that $d_i^{(\eta)} \geq 2i$. On the other hand, for the codes $C_i^{(\eta)\perp}$, we see that as i is incremented beginning with $i = 1$, the minimum distance remains fixed and equal to η until $i = \frac{\eta+1}{2}$. Incrementing i further, however, causes the minimum distance to start decreasing.

We also observe that for the two different η, the highest possible minimum distance achieved for the constructed quantum codes is $\eta - 1$, and occurs when either $i = \lfloor \frac{\eta+1}{2} \rfloor - 1$ or $j = \lfloor \frac{\eta+1}{2} \rfloor + 1$. In addition, if both $i = \lfloor \frac{\eta+1}{2} \rfloor - 1$ and $j = \lfloor \frac{\eta+1}{2} \rfloor + 1$ are satisfied, then the resulting code has the highest possible dimensionality (see for example the codes $\mathrm{CSS}\left(C_3^{(7)}, C_5^{(7)}\right)$ and $\mathrm{CSS}\left(C_5^{(7)\perp}, C_3^{(7)\perp}\right)$).

Generalizing these observations to larger prime η suggests the following heuristic design rules that, we strongly believe, will ensure the highest possible dimensionality and minimum distance for quantum codes constructed from diagonals on square arrays.

1. To achieve a quantum code with minimum distance d, we must choose prime η such that $d \leq \eta - 1$.

2. Given that condition 1 is satisfied, one can choose i to be the smallest integer that satisfies $2i \geq d$. This will ensure a minimum distance of $\geq d$.

3. Once η and i have been determined, we choose j such that $j > i$ and $i + j = \eta + 1$ to obtain the largest possible dimensionality. The quantum codes $\mathrm{CSS}\left(C_i^{(\eta)}, C_j^{(\eta)}\right)$ or $\mathrm{CSS}\left(C_j^{(\eta)\perp}, C_i^{(\eta)\perp}\right)$ will have a minimum distance of $\geq d$ and the highest dimensionality possible for that minimum distance.

4. If a quantum code rate higher than that specified by $\mathrm{CSS}\left(C_i^{(\eta)}, C_j^{(\eta)}\right)$ or $\mathrm{CSS}\left(C_j^{(\eta)\perp}, C_i^{(\eta)\perp}\right)$ is desired, we must choose successively higher prime η and go through steps 2 and 3 till the desired rate is achieved. Of course, the increase in rate will be necessarily accompanied by an increase in code length.

Here we reiterate that the stated design rules are strictly heuristic in nature and are based on the observations for specific values of η presented in this section.

We have reasonable confidence, based on our experience with this class of codes, to believe that the generalizations of these results to larger prime η will hold.

We can determine how the quantum codes presented in this section compare with known codes, by referring to the bounds (lower and upper) on the highest possible minimum distance achievable by quantum codes for specific n and k that are presented in [13]. Note that these bounds have been computed for quantum codes based on classical codes over GF(4). Nevertheless, one may use the bounds for the purposes of comparison[9]. For $n = 25$ and $k = 8$, the highest possible minimum distance is bounded between 4 and 7 [13]. Our $[25, 8, 4]$ code $\mathrm{CSS}\left(C_2^{(5)}, C_4^{(5)}\right)$ certainly compares reasonably well with this bound. Unfortunately, bounds for $n > 36$ are not available [30] for a direct comparison of parameters with our $[49, 12, 6]$ code $\mathrm{CSS}\left(C_3^{(7)}, C_5^{(7)}\right)$. However, from the table of bounds given in [30], we see that the highest possible minimum distance for a $n = 36$, $k = 12$ quantum code is lower bounded by 7. Our $[49, 12, 6]$ quantum code, although longer, has a minimum distance of only 6 and, hence, is much worse than the predicted highest possible minimum distance.

We can also compare our $[25, 8, 4]$ quantum code with the lower and upper bounds on the highest achievable minimum distance for CSS codes constructed

---

[9]Classical codes over GF(4) may be used to construct quantum codes [13]. The resulting quantum codes, however, are still binary in the sense that they are used for error correction in qubits (collection of binary quantum systems). Thus, one may make a valid comparison between a quantum code constructed from binary classical codes and a quantum code constructed from classical codes on GF(4). Of course, since GF(2) $\subset$ GF(4), classical codes over GF(4) offer many more possibilities as compared to classical binary codes. Consequently, it is fair to expect that better quantum codes may be constructed from classical codes over GF(4) as compared to quantum codes from classical binary codes.

from *weakly-dual* binary classical codes [30]. For $n = 25$ and $k = 7$, as well as $k = 9$ (note that $n = 25, k = 8$ is not possible using *weakly dual* binary codes), the highest possible minimum distance equals 4, which is also achieved by our code. Again, these bounds are not available for $n > 36$.

Next we discuss a related technique for the construction of *weakly dual* or *dual-containing* classical codes based on diagonals in $\eta \times \eta$ arrays, where η is even (also see [53]).

D.   Weakly-Dual Codes Based on Diagonals in Arrays with Even η

In constructing classical codes from diagonals based on $\eta \times \eta$ square arrays discussed earlier, η was chosen to be prime to ensure that the set of parity-check equations defined by the diagonals were concurrent or orthogonal on every bit position (i.e., to ensure that the bipartite graph of the code does not have 4-cycles), no matter which specific diagonal bundles were used. Choosing η to be odd also results in codes without 4-cycles provided certain specific diagonal sets are chosen to construct the H-matrix. Choosing η to be even is avoided, however, since such codes are prone to 4-cycles and always result in small minimum distances.

Absence of 4-cycles implies that the overlap between any two rows of the parity-check matrix is either 0 or 1. Thus, if we recall the conditions that the parity-check matrix must satisfy to ensure *weak duality*, we see that codes whose bipartite graphs do not have 4-cycles cannot be *weakly-dual*[10]. Thus, codes

---

[10]Here, we ignore the simplistic case where any two rows of the parity-check matrix have

based on diagonals in square arrays with η prime are never *weakly dual*.

Although not suitable for constructing good regular LDPC codes, square arrays with even η are potentially excellent candidates for constructing *weakly-dual* codes. First, in constructing PCEs from the diagonals of the array, every PCE has weight equal to η. Using arrays with even η ensures that the first of the conditions for constructing *weakly-dual* codes is satisfied. PCEs overlapping each other at an even number of positions and, hence, meeting the second condition, can also be guaranteed if the diagonal sets are appropriately chosen. Next, we briefly describe this second construction technique for quantum codes.

1. Construction of weakly dual codes on arrays with even η

Consider a classical code constructed on an $\eta \times \eta$ array with even $\eta \geq 4$. We choose the rows of the array (slope-0 diagonals) to constitute the first set of PCEs (called row PCEs). Since no two rows of the array intersect, the set of row PCEs have zero overlap. Once the rows of the array are chosen, one may not choose the columns (slope-∞ diagonals) to form a set of PCEs, since a column of the array will intersect any row of the array at exactly one position, thus resulting in an odd overlap. One may choose, however, the set of slope-s diagonals, where s is even: let us say $s = 2$. With a little thought, it becomes clear that any such slope-2 diagonal intersects alternate rows of the array at exactly two positions, while it does not intersect the other rows. Further, any two diagonals having the same slope never intersect. Thus, by using the slope-0 and slope-2 diagonal

zero overlap.

sets to form the PCEs, one can construct a parity-check matrix that represents a *weakly-dual* code. More generally, a *weakly-dual* code is assured so long as one uses diagonal sets with even slopes to construct the parity-check matrix.

For arbitrary square arrays with even η, one may view the code constructed from even-sloped diagonals as alternately arising from diagonals of any slope s (odd or even with $0 \leq s < \frac{\eta}{2}$) on an array with $\frac{\eta}{2}$ rows and η columns. This follows since the bipartite graph of the code with parity-check equations defined by the even-sloped diagonals contains two disconnected partitions[11]. The first partition consists of all elements on the even numbered rows and the parity-check equations that constrain only the elements in this partition, while the second partition includes all elements on the odd numbered rows and the parity-check equations that constrain only the elements in this partition. As such, one may discard all the elements on the odd numbered rows (or alternately all the elements on the even numbered rows) to give a $\frac{\eta}{2} \times \eta$ array.

We shall label such rectangular arrays as *type-R* to distinguish them from the square arrays. The process of contracting an $\eta \times \eta$ square array to the corresponding $\frac{\eta}{2} \times \eta$ *type-R* array is illustrated in Figure VII.1. The × symbols and the ○ symbols show a slope-2 and slope-4 (even-sloped) diagonal, respectively, on the square array. On the resultant *type-R* array, these symbols denote a slope-1 and slope-2 diagonal, respectively. Similarly, the *type-R* array for $\eta = 10$ is shown in Figure VII.2: the rows of the *type-R* array correspond to the set of slope-0 diagonals, the array elements indicated by the + symbol are constrained by one

---

[11]These two partitions are said to be disconnected since there exists no parity-check equation that causes an element from one of the partitions to interact with an element from the the other partition.

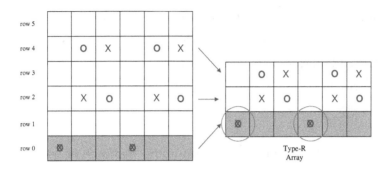

Figure VII.1: Contraction of a 6×6 square array to a 3×6 *type-R* array by discarding odd-numbered rows: (a) $\times$ symbols $\sim$ a slope-2 diagonal on the square array and, equivalently, a slope-1 diagonal on the *type-R* array, (b) $\circ$ symbols $\sim$ a slope-4 diagonal on the square array and, equivalently, a slope-2 diagonal on the *type-R* array, (c) shaded positions $\sim$ a slope-0 diagonal (row) on both arrays. The circled positions in the *type-R* array show an example of the bit-positions with value 1 in a codeword of Hamming weight-2.

of the $\frac{\eta}{2} = 5$ parallel slope-1 diagonal PCEs, while array elements with the $\times$ symbol are constrained by one of the $\frac{\eta}{2} = 5$ parallel slope-2 diagonal PCEs.

On a *type-R* array, we may use any number of slope-sets from the $\frac{\eta}{2}$ possible slopes to construct classical codes of length $\frac{\eta^2}{2}$ with different dimensionality. In every case, we are assured of having a *weakly-dual* code; the resulting classical code, however, has a large number of codewords of Hamming weight 2, which is not desirable, especially since these codewords also do not belong to the dual.

In Figure VII.1, the circled positions show an example of the bit-positions with value 1 in a codeword of Hamming weight-2. It is easily verified that all the PCEs are satisfied when the 1s are located at the two indicated positions. In fact, every codeword at Hamming weight-2 has a configuration similar to the codeword shown in Figure VII.1. Classical codes with a minimum distance of

2 are not very useful in constructing quantum codes; larger minimum distances are needed to construct useful quantum codes from such codes.

In the following subsection we discuss the different methods that we have investigated to successfully increase the minimum distance of codes on *type-R* arrays, while retaining their *weak duality*.

2. Methods to improve the minimum distance of weakly dual codes from diagonals on *type-R* arrays

For the *weakly dual* classical codes from *type-R* arrays to be useful in constructing quantum codes, we need to find ways to increase the minimum distance of these codes without disrupting their *weakly-dual* property.

One of the techniques involves additionally constraining every pair of consecutive columns on the *type-R* array to have even parity. We call such PCEs as *twin-column* PCEs.

In Figure VII.2 a *twin-column* PCE constraining the first two columns of the *type-R* array is represented by the shaded elements. It can easily be ascertained by observation that the introduction of the set of *twin-column* parities is sufficient to eliminate all codewords of Hamming weight-2, i.e., the codewords described by the configuration indicated by the circled elements in Figure VII.1. Further the structure of the *twin-column* parities ensures that *weak duality* of the code is retained, since such a parity has a weight of η (which is even) and overlaps any diagonal PCE exactly two times.

We have successfully constructed $\left(\frac{\eta^2}{2}, \left(\frac{\eta^2}{2} - \frac{3\eta}{2} + 2 \right) \right)$ *weakly-dual* classical codes with minimum-distance 4 using slope-0, slope-1, and *twin-column* parity sets on the $\frac{\eta}{2} \times \eta$ array for $\eta \in \{4, 6, 8, 10, 12, 14\}$. In general, we can show

Table VII.5

Classical *weakly dual* codes based on slope-0, slope-1, and *twin-column* parity sets on the $\frac{\eta}{2} \times \eta$ array for $\eta \in \{4, 6, 8, 10, 12, 14\}$ and the corresponding CSS quantum codes.

η	Classical code (n, k, d)	Quantum code $[n, k, d]$
4	$(8, 4, 4)$	$[8, 0, 4]$
6	$(18, 11, 4)$	$[18, 4, 4]$
8	$(32, 22, 4)$	$[32, 12, 4]$
10	$(50, 37, 4)$	$[50, 24, 4]$
12	$(72, 56, 4)$	$[72, 40, 4]$
14	$(98, 79, 4)$	$[98, 60, 4]$

that these results hold for all even $\eta \geq 4$. These *weakly-dual* classical codes in turn give us $\left[\frac{\eta^2}{2}, \left(\frac{\eta^2}{2} - 3\eta + 4\right)\right]$ CSS codes with minimum-distance 4 and single error-correcting capability. The values of η and the corresponding classical and quantum code parameters for codes constructed from slope-0, slope-1, and *twin-column* parity are listed in Table VII.5.

We also discovered that, for some values of even η, it is possible to obtain an increase from 4 to 6 in the minimum distance of the classical code (and hence the CSS code). This is achieved by further constraining the bit-positions in the classical code via a set of so called *mirrored diagonals* of slope-*s*, followed by removing the codeword bit-positions on a pair of columns on the *type-R* array.

A *mirrored* slope-*s* diagonal PCE is the same as a slope-*s* diagonal PCE on the left-half of the *type-R* array. The positions that are checked by the PCE on the right-half of the *type-R* array are determined by simply reflecting the checked positions on the left-half about the vertical that splits the *type-R* array into two halves (see Figure VII.2, where the array elements with the ○ symbol correspond to those elements constrained by a *mirrored* slope-2 diagonal PCE).

We emphasize that using the set of *mirrored* diagonal parities of a specific slope instead of the regular diagonal parities of that slope does not eliminate all codewords at Hamming weight-4. However, it certainly results in a significant reduction in the number of codewords at Hamming weight-4. As an example, for $\eta = 10$, enforcing slope-0 (row), slope-1, *twin-column*, and slope-2 diagonal parities on the corresponding *type-R* array resulted in an $(n = 50, k = 33)$ code with 50 codewords at Hamming weight-4. On the other hand, enforcing slope-0 (row), slope-1, *twin-column*, and *mirrored* slope-2 diagonal parities on the corresponding *type-R* array resulted in an $(n = 50, k = 33)$ code with only 15 codewords at Hamming weight-4. The removal of codeword positions on a pair of columns is further required in order to eliminate the few remaining codewords at Hamming weight-4.

The process of removing positions from a specific column of the array destroys the self-orthogonality of the *twin-column* parity-check equation that originally constrained the removed column positions; the orthogonality of this PCE with respect to the other PCEs also is destroyed. Since we remove the positions on a pair of columns on the array, *weak duality* can be immediately restored by combining the remnants of the two *twin-column* PCEs that are each disrupted by the elimination of one of their columns.

Specifically, for the *type-R* array with $\eta = 10$ (see Figure VII.2), we used the technique just described to construct a *weakly-dual* $(40, 24, 6)$ classical code

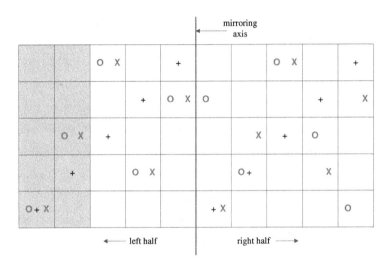

Figure VII.2: The 5 × 10 *type-R* array ($\eta = 10$) with: (a) rows of the array ∼ slope-0 diagonals, (b) + symbols ∼ a slope-1 diagonal, (c) × symbols ∼ a slope-2 diagonal, (d) ∘ symbols ∼ a *mirrored* slope-2 diagonal, (e) shaded positions ∼ a *twin column* parity.

that, in turn, gives a $[40, 8, 6]$ CSS quantum code. The parity-check matrix was constructed by using the slope-0, slope-1, *mirrored* slope-2 diagonals, and the *twin-column* parity. This was followed by discarding all the codeword bit-positions that belonged to the third and eighth columns (the columns are indexed from left to right beginning with 1). These bit-positions were removed because it was observed that every codeword at Hamming weight-4 had a one position in at least one of the codeword bit-positions on these two columns. The removal of these codeword bit-positions, thus, ensured that there were no more Hamming weight-4 codewords in the shortened code.

The shortened code was not *weakly dual*, however, thanks to the fact that the *twin-column* PCEs, involving the third and fourth columns and the seventh and eighth columns, were each disrupted and now had only five participating codeword bit-positions. *Weak-duality* was easily restored by combining the two disrupted *twin-column* PCEs into a single new PCE.

We have observed that $\eta/2$ prime is a necessary requirement for this technique to be effective in raising the minimum distance. We note that a cyclic quantum code with the same parameters has been reported previously [30], but the reported construction is based on an additive self-dual code over GF(4).

Similarly, we also were able to construct a $(84, 54, 8)$ *weakly-dual* classical code from the $\eta = 14$ *type-R* array by using slope-0, slope-1, slope-3, and *mirrored* slope-2 diagonals, and the *twin-column* PCEs. This, in turn, gives us a $[84, 24, 8]$ CSS quantum code capable of correcting 3 errors.

As noted previously, bounds on the highest achievable minimum distance for CSS codes from *weakly dual* binary classical codes are not provided for $n > 36$. For $n = 36$ and $k = 4, 6, 8,$ and 10, however, the highest achievable minimum distance is lower bounded by 6 [30]. The $[40, 8, 6]$ quantum code we constructed compares reasonably well with these bounds.

The $[84, 24, 8]$ quantum code we constructed is much too long to hazard an educated guess for bounds on the highest achievable minimum distance. We note that a $[64, 20, 8]$ quantum code [61], however, has been constructed by applying the CSS technique to the class of Reed-Muller codes [44]. This code has the same minimum distance as our $[84, 24, 8]$ code, but is shorter and has higher rate.

Finally, it is worthwhile pointing out that the parity-check matrices of the classical codes constructed via the method described in this dissertation are sparse with density $= \frac{\gamma}{(\gamma-1)\frac{\eta}{2}+\left(\frac{\eta}{2}-1\right)} \approx \frac{2}{\eta}$, and, importantly, they also are column regular with column weight γ, where γ is the total number of sets of PCEs (one set per slope bundle) used in constructing $\mathbf{H}$ ($\gamma = 4$ for the $(40, 24, 6)$ code and $\gamma = 5$ for the $(84, 54, 8)$ code). The decoding operation of CSS quantum codes requires the measurement of syndromes and, for the quantum codes constructed, each qubit is required in the computation of only 2γ syndromes, which is a small number if γ is small. Having a small number of interactions per qubit in the decoder is indeed desirable, since the probability of corrupting the state of a qubit increases with the number of interactions [49].

E.   Conclusions

1.   Summary and conclusions

In this chapter we discussed the construction of *quantum error correction* codes from the class of codes based on diagonals in square arrays with η prime via the Calderbank-Shor-Steane construction technique. We explicitly constructed such quantum codes from diagonals based on 5×5 and 7×7 square arrays, and based on observing the properties of these quantum codes, we presented some heuristic design rules for the construction of such quantum codes having a specific minimum distance and the highest possible dimensionality.

We also presented a technique to construct binary *weakly-dual* classical forward error-correction (FEC) codes based on diagonals of different slopes on a rectangular $\frac{\eta}{2} \times \eta$ array (*type-R* array) with η even. Some methods to increase the minimum distance of these classical codes also were discussed. The quantum codes obtained from both construction techniques were compared with some known quantum codes of similar dimensions and/or with bounds (for quantum codes constructed from either classical *weakly-dual* binary codes or classical codes over GF(4)) on the highest possible minimum distance for the specified length and dimensionality.

We conclude that the classical codes based on diagonals in square arrays lend themselves admirably to the construction of quantum codes via the CSS construction technique. The set of rules-of-thumb proposed for constructing such quantum codes with specified minimum distances is strictly heuristic and no formal results have been proven in this context. We also conclude that one can use the structure of even-sloped diagonals on $\eta \times \eta$ arrays with η even to construct *weakly dual* classical codes. These codes, however, have a small minimum distance of two. Techniques such as the addition of *twin-column* parity and *mirrored* slope PCEs succeed in increasing the minimum distance of such codes without disrupting their *weak duality*. The resultant classical codes also have the advantage of having sparse and column-regular parity-check matrices, which implies that the qubits of the quantum code are required to participate in only a few syndrome computations in the decoding process.

The two longer quantum codes we constructed (the $[49, 12, 6]$ and $[84, 24, 8]$ codes) do not compare as well with some other known quantum codes in terms of their $[n, k, d]$ parameters. The minimum distances of two of the shorter quantum codes we constructed (the $[25, 8, 4]$ and $[40, 8, 6]$ codes), however, compare favorably with the minimum distances of other known quantum codes and with known bounds on the highest achievable minimum distance for specific n and k.

2. Future research

A large number of interesting research possibilities arise from the material presented in this chapter.

1. We would like to obtain mathematical expressions and/or bounds for some of the values of the parameters (dimensionality and minimum-distance) for the classical codes constructed from diagonals on square $\eta \times \eta$ arrays with η prime. These results, in turn, will help us exactly determine the parameters of the quantum codes that may be constructed from such classical codes.

2. It also would be useful to obtain the minimum distances of all the classical codes possible and their duals, based on diagonals in $\eta \times \eta$ arrays, for $\eta = 11$ and 13 (the next two primes after 5 and 7). Note, however, that the task of numerically computing the minimum distance of codes becomes exceedingly time consuming very rapidly.

3. Since the classical codes based on diagonals in square arrays may be equivalently obtained from other algebraic/geometric structures such as Euclidean geometries [44] and partial BIBDs [70] (see Chapter II), it is rea-

sonable to expect that these other classical code construction techniques also yield a class of classical codes that can be used to construct quantum codes via the CSS technique in a manner similar to that presented in this chapter. We recommend an investigation of these construction techniques to determine if quantum codes with superior $[n, k, d_{\min}]$ parameters are possible.

4. We need to investigate methods to further enhance the minimum distance of classical codes based on the *type-R* arrays. An example of a possible enhancement is the introduction of permutations on the columns of the *type-R* array instead of mirroring of the slope. Such enhancements may altogether eliminate all the codewords at the lowest Hamming weight, rather than just reducing the number of codewords at the lowest Hamming weight. The process of eliminating the codeword bit-positions, thus, may be rendered unnecessary.

5. A decoder circuit for CSS quantum codes may be easily constructed once the *check matrix* [49] of the code is known. The conventional decoding scheme for quantum codes involves measuring the qubits of the quantum code with respect to measurement operators constructed based on the rows of the *check matrix*; each row of the check matrix corresponds to a specific measurement operator. The measurement outcome of the qubits of the quantum code with respect to a measurement operator is called the syndrome. The set of syndromes corresponding to the rows of the *check matrix* are then used to perform appropriate unitary transformations on the qubits of the quantum code to perform error-correction.

It also may be possible to devise low-complexity decoding schemes, however, for decoding the quantum codes resulting from our classical codes, much in the same way as such decoding schemes have been found for classical codes (such as LDPC codes). We must be mindful, of course, of the impact and limitations of quantum measurements in implementing quantum code decoders. In particular, the potentially adverse impact that repeated measurements (and, hence, interactions with the environment) can have on the state of a qubit makes it advantageous to have qubits of the quantum code participating in as few syndrome measurements as possible. The minimization of syndrome measurements per qubit may be used as a criterion to design quantum codes, and we suggest an investigation in this direction.

Chapter VIII.

SUMMARY AND CONCLUSIONS

This dissertation is concerned with the class of row-column-diagonal (RCD) codes parameterized by a prime integer η. RCD codes are a class of regular low-density parity-check codes with fixed minimum distance = 6 that can achieve very high code rates with increasing η. This class of codes can be decoded via a range of low-complexity decoding algorithms that provide a variety of trade-offs between performance and decoding complexity and speed. This class of codes is thoroughly analyzed with regards to its properties, and decoding and performance evaluation on various channels such as the binary symmetric channel (BSC), binary symmetric channel with erasures (BSC/E), and the binary input continuous output channel with additive white Gaussian noise.

In this, the final chapter of the dissertation, we summarize our research, discuss our conclusions, and provide directions for future research.

A. Summary

In the first chapter of the dissertation, we provided an introduction to the dissertation and motivated the study of the class of RCD codes.

289

In the second chapter, we introduced the class of regular LDPC codes constructed on an $\eta \times \eta$ square array, where η is prime, with γ slope-q diagonal bundles representing the parity-check equations. The RCD codes are a special case of such codes with $\gamma = 3$. The name RCD stems from the fact that the three slope-q diagonal bundles used in the construction of RCD codes are the rows (slope-0 diagonal bundle), columns (slope-∞ diagonal bundle), and the main diagonals (slope-1 diagonal bundle). In this chapter, we established an equivalence between the Gallager LDPC codes constructed from diagonals on the square array and the Gallager LDPC codes constructed based on Euclidean Geometries (EG) with parameters $m = 2, s = 1, \eta$, and γ. An equivalence between the former class of codes and Gallager LDPC codes based on (η, γ) lattice partial-balanced incomplete-block designs (PBIBDs) was also similarly established. Some of the important properties of RCD codes were also discussed in this chapter.

In the third chapter, we analyzed the decoding of RCD codes via the bit-flipping algorithm and evaluated the performance of RCD codes on a binary symmetric channel. For RCD codes, we showed that the BFA is capable of correcting all error patterns up to Hamming weight equal to $\lfloor \frac{d_{\min}-1}{2} \rfloor = 2$, where $d_{\min} = 6$ is the minimum distance of the RCD code. Thus, for the RCD code, we showed that the BFA is capable of performing at least as well as a bounded-distance decoder. We then analyzed the decoding of errors patterns of Hamming weight-3 via the BFA. A classification scheme was developed for the 3-error patterns based on the interactions, via parity-check equations (PCEs), of the three positions in error. Combinatorial expressions were obtained for the number of error patterns belonging to each of the classes. Important theoretical results characterizing the decoding under BFA of the various classes and sub-classes of 3-error patterns were proved. These theoretical results were then used to construct a lower bound on the number of 3-error patterns that decode successfully to the all-zeros codeword.

We then presented experimental results for the exhaustive generation, classification, and BFA decoding of 3-error patterns for RCD codes with $5 \leq \eta \leq 23$. Similarly, a classification scheme was also provided for 4-error patterns, and experimental results based on the exhaustive generation, classification, and BFA decoding of 4-error patterns for RCD codes with $5 \leq \eta \leq 17$ were also obtained.

Upper and lower bounds on the word-error rate (WER) performance of certain RCD codes on a binary symmetric channel were computed, based on the theoretical results on the number of 3-error patterns that decode successfully to the all-zeros codeword. WER upper- and lower-bounds from exhaustive searches of 3-error patterns only, as well as of 3- and 4-error patterns, also were computed for those values of η where such computations were possible. For those ηs where results from an exhaustive search were available, the bounds from the exhaustive search were compared with the theoretical upper and lower WER bounds and found to agree. All the computed bounds also were compared with standard Monte Carlo simulations results of WER and were found to agree well.

In Chapter IV, we addressed the decoding of codes in general, and RCD codes in particular, on the binary symmetric channel with erasures (BSC/E). For the purpose of our analysis, we used a simple extension of a hard-decision decoder to decode on the BSC/E. We have presented a comparison of WER performance improvement for codes with a given n and d_{min}, and under extended bounded distance decoding (e-BDD), in going from a BSC to a BSC/E channel model derived using a BPSK-AWGN system with decision thresholds at $\pm t$. The performance of our RCD codes was of particular interest in this study and motivated the specific choices of code parameters that were studied. The plots obtained enable us to determine the minimum WER threshold $t^*(\text{SNR})$ under e-BDD for the BSC/E for a given code and desired channel signal-to-noise ratio (CSNR).

The performance of RCD codes on the BSC/E also was analyzed under decoding via the extension to the BFA (e-BFA). Some bounds on e-BFA performance for RCD codes were derived, and plots of e-BFA performance were presented for the $\eta = 23$ RCD code. The simulation results and bounds for e-BFA are compared also with simulation results for decoding on the BSC/E using a different message passing decoder.

The soft-decision iterative-decoding performance of some RCD codes was evaluated in the fifth chapter of the dissertation. We presented performance curves in the form of WER and BER vs. E_b/N_0 for three different RCD codes ($\eta = 17$, $\eta = 23$, and $\eta = 37$), where the dual adaptive importance sampling (DAIS) technique based on multicanonical Monte Carlo (MMC) simulations was employed to evaluate code WER and BER performance. The DAIS technique is employed because it allows us to calculate very low WERs and BERs: far below the capabilities of standard Monte Carlo (MC) estimation techniques. The BER/WER results from the DAIS evaluation are compared with corresponding results obtained via standard MC simulation, and with the union bound on optimal maximum-likelihood decoding performance. The resulting comparisons enable us to ascertain the validity of DAIS as a WER/BER estimation technique and, also, to explain the relationship between the WER performance curve of the RCD code and the union bound approximation, in terms of the relative contribution of the two events that constitute errors: decoder failures and decoding-to-wrong codewords.

Chapter VI of the dissertation is concerned with an importance sampling technique for hard-decision decoding performance of codes that was inspired in part by the evaluation of error patterns of specific weights for RCD codes under BFA decoding. In this chapter, we presented the concept of a hard-decision FEC decoder output's SNR-invariance to the binary symmetric channel (BSC) transition probability p (defined by the SNR) given a received length-n binary word. We then used this SNR-invariance concept to develop an importance sampling (IS) technique (denoted invariance importance sampling (IIS)) for hard-decision decoding performance evaluation of binary linear block codes. The equations describing IIS were presented and examples demonstrating the IIS technique were provided. An analysis of the IIS estimator accuracy was provided in terms of its relative bias and variance. We then presented results of IIS simulations performed under BFA decoding for the $n = 255$, $k = 175$ Euclidean geometry FEC code, the $n = 256$, $k = 81$ 4-dimensional single-parity-check product FEC code, and the $\eta = 17$ and $\eta = 37$ RCD codes. We also evaluated the accuracy of the IIS estimator for each code.

In Chapter VII we discussed the application of the codes based on diagonals in square arrays to construct codes for performing error correction in the quantum domain (quantum codes). Specifically, we discuss the construction of *quantum error correction* codes from the class of classical codes based on diagonals in square arrays with η prime via the Calderbank-Shor-Steane construction technique. We explicitly constructed such quantum codes from diagonals based on 5×5 and 7×7 square arrays; based on observing the properties of these quantum codes, we presented some heuristic design rules for the construction of such quantum codes that have a specific minimum distance and the highest possible dimensionality. We also presented a technique to construct binary *weakly-dual* classical codes based on diagonals of different slopes on a rectangular $\frac{\eta}{2} \times \eta$ array (*type-R* array) with η even. Some methods to increase the minimum distance of such codes were also discussed. Quantum codes obtained from both construction techniques also were compared with some known quantum codes of similar dimensions, or with bounds (for quantum codes constructed from either classical binary codes or classical codes over GF(4)) on the highest possible minimum distance for the specified length and dimensionality.

B. Conclusions

The diagonals of various slopes on $\eta \times \eta$ square arrays provide a rich structure for the construction of binary forward error correction (FEC) codes as demonstrated in this dissertation. The row-column diagonal (RCD) codes with prime η are a special case of this class of codes, where three diagonal bundles, the rows, the columns, and the slope-1 diagonals, are used to construct the parity-check matrix of the code. Thus the RCD codes are a class of regular LDPC codes with column weight $\gamma = 3$.

The RCD codes may be viewed as an intermediate code between the two-dimensional single-parity check product (SPCP) code, which has minimum distance 4 and column weight $\gamma = 2$ [12], and the three-dimensional SPCP code, which has a minimum distance of 8 and column weight $\gamma = 3$ [12]. When constructed on multi-dimensional square arrays of side η (for any η), the 3D SPCP codes require a length of η^3 and a code rate $(\eta - 1)^3 / \eta^3$ to achieve a minimum distance of 8. This represents a significant expansion in the length of these codes and reduction in code rate as compared to the 2D SPCP codes that have length η^2 and rate $(\eta - 1)^2 / \eta^2$. An RCD code for a specific η has the corresponding 2D SPCP code as a sub-code. For prime η, the RCD code achieves a happy balance between the 2D SPCP code and the 3D SPCP code, by giving a stronger code than the 2D SPCP code at the cost of reduced rate, but without the rapid increase in code length and reduction in code rate that is associated with the 3D SPCP code.

Based on the discussion in Chapter II, we conclude that the codes constructed based on a $\eta \times \eta$ square array, where η is prime, with γ slope-q diagonal bundles representing the parity-check equations, is equivalent to Gallager LDPC codes constructed based on Euclidean Geometries (EG) with parameters $m = 2, s = 1, \eta$, and γ, and also Gallager LDPC codes based on (η, γ) lattice PBIBDs. Thus, we conclude that Gallager LDPC codes based on either of the three different construction techniques discussed eventually lead to exactly the same code.

We believe that amongst the three construction techniques, the technique based on diagonals in square arrays is the simplest to visualize, understand, and implement. Further, since the elements of the square grid represents the codeword elements, this approach to characterize the code construction also proves to be helpful in the exact evaluation of the weight enumerator function (WEF) of the code, especially when $\gamma = 3$, as demonstrated in [55]. Constructing codes using lines of Euclidean geometries involves performing a lot more mathematical computations as compared to diagonals on square arrays. This construction technique is mathematically and structurally richer, however, since one can construct Gallager LDPC codes on an m-dimensional grid whose side is not just a prime, but also a prime power. The construction technique based on lattice PBIBDs is similar to the technique based on diagonals in square arrays in terms of the computations required to generate the parity-check matrix. Further, the family of codes generated using lattice PBIBDs (we are specifically referring to lattices with prime η and blocks constructed from parallel lines in the lattice) offers exactly the same set of code parameters as the family of codes generated using diagonals on square arrays. However, the characterization of the code construction from lattice PBIBDs may not be as helpful as that from diagonals on arrays in the exact evaluation of the weight enumerator function (WEF) of the code when $\gamma = 3$.

The RCD codes constructed on $\eta \times \eta$ arrays have code length $n = \eta^2$ and do not have cycles of length 4 in their bipartite graphs. Having η prime (or, more generally, odd) ensures that the code does not contain codewords of Hamming weight 4. Also for η odd, RCD codes have dimensionality $k = (\eta - 1)(\eta - 2)$, and thus, their code rate is $r = \frac{k}{n} = \frac{(\eta-1)(\eta-2)}{\eta^2}$, which approaches unity as $\eta \to \infty$.

The class of RCD codes are a family of LDPC codes that are capable of achieving high code rates. Thanks to their regular LDPC structure, these codes also lend themselves to an array of decoding schemes that provide a variety of trade-offs in terms of decoder performance on the one hand, and decoder complexity on the other.

Further, the simple and elegant structure of the parity-check equations makes a combinatorial analysis of such codes possible in order to compute its weight enumerator function (WEF) [55], or to study its decoding under certain specific decoding algorithms.

From the results presented in Chapter III, we conclude that the bit-flipping algorithm is capable of correcting a large number of error patterns patterns beyond the bounded-distance error-correcting capability (of 2) for RCD codes. Thus, the BFA is at least as good as a bounded distance decoder (BDD) and, in fact, much better. The results presented for BFA decoding of RCD codes in this chapter show that an improvement of about 0.5 dB in code performance is obtained by using the BFA instead of the BDD for the RCD codes considered.

We conclude that the classification scheme for 3- and 4-error patterns presented in this chapter is exhaustive. However, extending this classification to higher-weight error patterns results in a rapid increase in the complexity of the process due to the progressive larger number of classes and sub-classes. Also, we conclude that a theoretical/combinatorial analysis of decoding of 3-error patterns of RCD codes is feasible and enables us to compute accurate lower and upper bounds on the word-error rate (WER) performance of the BFA decoder for RCD codes. These performance bounds are tight at high signal-to-noise ratio (SNR) and can be computed for the entire class of RCD codes without resorting to time-consuming simulations.

In Chapter IV, we focus on code performance on the binary symmetric channel with erasures (BSC/E), the simplest example of a channel that incorporates soft-information. For a variety of code parameters considered, which include some RCD code parameters, we can conclude that substantial gains in performance – either a reduction in WER for a given channel signal-to-noise ratio (CSNR) or a reduction in CSNR to achieve the same WER – can be obtained, especially at higher CSNR values, by employing two thresholds instead of one threshold for the BPSK-AWGN system with the threshold chosen to minimize the WER. All these results are based on a decoding algorithm that offers a simple extension to any hard-decision decoder for the BSC.

At a specific CSNR, these improvements increase with increasing $d_{\min}$ for a fixed length, but decrease with increasing length for a fixed $d_{\min}$, suggesting that the stronger codes benefit more from softer decoding. On the other hand, at a specific WER, the coding gain in dB decreases with n for fixed $d_{\min}$, and decreases with $d_{\min}$ for fixed n when the parity of $d_{\min}$ (even or odd) is kept fixed. From this viewpoint, thus, the stronger codes benefit lesser from decoding using erasures. Nevertheless, the coding gains obtained for the codes considered in this dissertation are significant to merit the use of a BSC/E instead of a BSC.

A distinct difference is observed between codes with odd and even $d_{\min}$ in the behavior of WER as a function of the BSC/E threshold t, and of the optimal BSC/E threshold t^* as a function of CSNR. Codes with even $d_{\min}$ such as RCD codes have a larger improvement in performance in going to a BSC/E from a BSC as opposed to codes with odd $d_{\min}$. This is primarily because the ability of even $d_{\min}$ codes to correct an extra erasure that is not utilized on the BSC becomes available on the BSC/E.

For a collection of codes on the BSC/E with the same $d_{\min}$ but different lengths, under the assumption that each code operates with the optimal threshold, each code is optimal for some sub-interval of E_b/N_0. Further, in such a case, for every E_b/N_0 value one can find a code that provides the best WER performance among codes in that collection. For codes with the same length but different even $d_{\min}$, the WER-minimizing thresholds are seen to converge to a non-zero value at very low CSNR. On the other hand, for codes with odd $d_{\min}$, the best performance at very low CSNR is achieved by setting $t = t^* = 0$ (i.e., by employing the BSC).

Based on the discussion presented, we conclude that, in general, the existence of a local minima for the WER of the e-BDD as a function of $t \geq 0$ is not guaranteed when $d_{\min}$ is odd. Furthermore, even if a local minimum exists, it need not be a global minimum. Thus, analytical minimization of the WER of the e-BDD as a function of $t \geq 0$ will require constrained optimization techniques. In fact, in all examples of codes with even $d_{\min}$ considered (which includes the RCD codes), we have always observed the existence of $t^* \neq 0$. Conclusive proof of this fact for all even $d_{\min}$, however, remains to be demonstrated.

For the specific case of RCD codes, one can perform a combinatorial analysis of e-error and f-erasure patterns that satisfy $2e + f = 6$ to compute a lower bound on the WER of RCD codes using the extended-BFA (e-BFA) algorithm on the BSC/E and assuming the transmission of the all-zeros codeword. The performance of the e-BFA under the assumption of transmitting the all-zeros codeword, however, is not an accurate measure of true e-BFA performance (i.e., with transmission of randomly-chosen codewords). In fact, the actual code performance is worse than the performance observed when only the all-zeros codeword is transmitted; in the latter case the procedure of setting all the erasures to zeros and ones, respectively, at the input of the two constituent decoders (CD) biases the e-BFA towards choosing the output of the first CD (the all-zeros codeword) over choosing the output of the second CD. Thus, the lower bound on code WER, obtained by analyzing performance under all-zeros codeword transmission, still serves as a lower bound on the code WER in general. This bound, however, is not tight at high SNR.

We also conclude that for the $\eta = 23$ RCD code, a coding gain of about 0.8dB may be obtained at a WER of 10^{-10} in communicating via a BSC/E with $t = 0.116$ and e-BFA decoding, instead of communicating via the BSC with BFA decoding. Moreover, this coding gain increases as WER decreases. Further e-BFA decoding may provide better performance than that provided by a message passing decoder (algorithm E of [65]) for the BSC/E.

Chapter V is devoted to the newly developed *dual-adaptive importance sampling* (DAIS) [33] technique evaluation of RCD codes under soft-decision iterative decoding via the sum-product algorithm (SPA).

Using DAIS for the RCD code with $\eta = 37$ ($n = 1369$, $k = 1260$), we were successful in evaluating WERs down to $\approx 10^{-10}$, while for the RCD codes with $\eta = 17$ ($n = 289$, $k = 240$) and $\eta = 23$ ($n = 529$, $k = 462$), respectively, we succeeded in estimating WERs to values lower than 10^{-15}. For all codes, the WER and BER results obtained via DAIS were seen to be consistent with standard Monte Carlo (MC) simulations and the union bound on maximum-likelihood decoding [23]. In particular, all DAIS WERs fell within the 99% confidence intervals of the corresponding standard MC WERs, wherever both DAIS and standard MC estimates were computed. Further all the DAIS results presented in this dissertation were obtained in a reasonable time frame of a few hundreds of hours at most (the maximum simulation time taken so far has been 840 hours for, not surprisingly, the $\eta = 37$ RCD code at 8dB). From all these results, we can conclude that DAIS may be used to evaluate code performance down to extremely low values of BER and WER in practical time frames.

Using the performance results obtained via DAIS, we were able to explain the relationship between the WER performance curve of the code and the union bound approximation at different E_b/N_0 in terms of the relative contributions of the two events that constitute errors: decoder failures and decoding-to-wrong codewords. At the low and high E_b/N_0 values, the dominance of decoder failures over decoding-to-wrong codewords leads to a wide separation between the WER performance and the union bound approximation. On the other hand, in the moderate E_b/N_0 region, the significantly higher error contribution from decoding-to-wrong codewords results in the close proximity of the WER performance to the union bound. This also corresponds to the region where the SPA decoder most closely approximates the optimal ML decoder[1] Rather interestingly, at much higher E_b/N_0, the ability of the SPA to closely approximate the ML decoder diminishes.

The performance curves of the RCD codes are observed to fall off with a rather gentle slope on account of the relatively low value of the RCD code minimum-distance ($d_{min} = 6$). Also, the $\eta = 17$ and $\eta = 23$ RCD code performance curves do not exhibit a waterfall region, most likely since the codes are not sufficiently long; this also could be due to the relatively low minimum distance of the codes, which results in the code performance curve approaching the union bound at a relative high WER of about 10^{-3} to 10^{-4}. A weak waterfall region is visible, however, for the $\eta = 37$ RCD code. One would expect the waterfall region to become increasingly prominent only as η increases.

Another interesting conclusion can be drawn by comparing the value of E_b/N_0 that achieves a specified BER for a specific RCD code with the Shannon-limit for the corresponding code rate. We specifically consider the Shannon limit for

---

[1] The ML decoder always decodes to a valid codeword and, hence, does not lead to decoder failures.

a binary-input continuous-valued output channel model (which agrees with the BPSK-AWGN model that we have assumed). For such a channel, the capacity is given by equation (3.97) in [3], and the resulting Shannon limits at code-rates of 0.83 ($\eta = 17$ RCD code), 0.87 ($\eta = 23$ RCD code), and 0.92 ($\eta = 37$ RCD code) are 2.31dB, 2.79dB, and 3.53dB, respectively. Comparing these values of E_b/N_0 with the E_b/N_0 values that achieve BER $\approx 10^{-10}$ for the corresponding RCD codes, which are 7.2dB, 7.1dB, and 7.0dB, respectively, clearly indicates that the gap between the codes' E_b/N_0 values and the corresponding Shannon-limit decreases with increasing η. Thus longer RCD codes also should have performance approaching their Shannon-limits, which reflects their categorization as weakly random-like codes. The extension of this fact to larger η is further reinforced by our observation that the BER union bounds of the RCD codes when plotted for different η do not show as much horizontal-shift variation (see the curves in Figure V.10) as the Shannon-limits (see the lines in Figure V.10) for the same values of η.

Using the BER union bound for the RCD codes as a reference for code performance, we conclude that for a desired BER there is a range of values of η for which the corresponding RCD codes provide the lowest E_b/N_0. For a BER of 10^{-15} the range of η turns out to be approximately between 37 and 101. It is also clear that at a given E_b/N_0, the performance improvement that accrues due to increasing code-rate as η increases is offset by the loss in performance due to a corresponding increase in the term $\frac{A_6}{n}$ (see equation (V.3)), which is proportional to η^2.

Based on the performance curves of the three RCD codes, and intuitively, one would expect that for SPA decoding decoder failures will progressively become the more dominant contributor to the probability-of-error over a larger range of E_b/N_0 values as η increases. An indication in this regard is obtained by observing that the SPA WER performance is close to the corresponding union bound over a larger range of E_b/N_0 values for the $\eta = 17$ RCD code than for the $\eta = 37$ RCD code.

Finally, it is worth pointing out that the decoding complexity of the SPA per iteration is proportional to $6n\gamma$ [47], where n is the code length and γ is the column weight of the regular LDPC code. RCD codes always have $\gamma = 3$. Thus, for a specific code length, we conclude that the SPA decoder for RCD codes has the lowest complexity among decoders for regular LDPC codes of that length. This feature makes RCD codes particularly desirable for high-rate low decoding-complexity applications.

The exposition in Chapter VI deals with the SNR-invariant importance sampling (IIS) technique that we developed, which may be viewed as an application of the *method of stratification* [42] to the problem of evaluating performance of FEC codes under bounded distance decoding. We conclude that it is possible to use the proposed IIS technique to accurately (with negligible relative bias and small variance) evaluate word-error rate (WER) and bit-error rate (BER) to arbitrarily low values for the chosen codes. In fact, a significant advantage of the IIS technique is that under appropriate conditions, such as small-to-moderate code lengths or when knowledge of the error correcting capability of the decoder is known, the results obtained from a single IIS simulation are sufficient for accurately evaluating WERs and BERs for a range of SNRs, where the upper end of the range may often be any arbitrarily chosen high SNR value.

We also conclude that even without definitively knowing whether the decoder can correct all error-patterns at certain Hamming weights, it is still possible for the IIS-estimator to improve the range of SNR and WER for which we have reliable estimates, as compared to the standard MC estimator with the same total number of transmitted words. Further, in such a case, the probability estimates obtained via IIS always serve as a lower bound on the true probability values.

The performance results obtained by evaluating RCD codes via IIS under BFA decoding on the BSC agree exceedingly well with the theoretical bounds as well as standard MC simulation results. The IIS simulation results all were obtained in reasonable time frames on commonly available current technology workstations (e.g., a dedicated Pentium-4 2.4 GHz 2GB RAM machine). The total number of decoded words and the approximate times taken for the various codes are provided in Table VI.9. These simulation times are significantly lower than what would be required by standard Monte Carlo simulations to estimate low WERs. For example, a standard Monte Carlo simulation for estimating the low WER of the order of 10^{-12} is likely to require approximately hundreds of years on a similar computer for all codes considered. For lower WERs, the time requirements of standard Monte Carlo are only higher.

For a particular RCD code, the vertical separation between the BFD WER curve and the bounded-distance decoder (BDD) WER curve at a specific high E_b/N_0 corresponds to a factor that equals the raw WER at Hamming weight 3 for that code. Since the raw WER at Hamming weight 3 decreases for increasing η for the class of RCD codes, we conclude that at a high value of E_b/N_0 the WER improvement obtained by BFD decoding over BDD decoding for an RCD code is larger for larger η. In other words, RCD codes with larger η derive more benefit at high E_b/N_0 from using the BFD instead of the BDD.

From the research presented in Chapter VII, we conclude that the codes based on diagonals in square arrays lend themselves admirably to the construction of quantum codes via the CSS construction technique. The set of rules-of-thumb proposed for constructing such quantum codes with specified minimum distances is strictly heuristic and no formal results have been proven in this context. We also conclude that one can use the structure of even-sloped diagonals on $\eta \times \eta$ arrays with η even to construct *weakly dual* classical codes. These codes, however, have a small minimum distance of two. Techniques such as the addition of *twin-column* parity and *mirrored* slope parities succeed in increasing the minimum distance of such codes without disrupting their *weak duality*. The two longer quantum codes we constructed (the $[49, 12, 6]$ and $[84, 24, 8]$ codes) do not compare as well with some other known quantum codes in terms of their $[n, k, d]$ parameters. The minimum distances of two of the shorter quantum codes we constructed (the $[25, 8, 4]$ and $[40, 8, 6]$ codes), however, compare favorably with the minimum distances of other known quantum codes and with known bounds on the highest achievable minimum distance for specific n and k.

In closing this section, we point out that RCD codes may not be the best (in terms of error correcting performance) available codes for a particular length and code rate; in fact, BCH codes of lengths and rates comparable to RCD codes achieve much higher error correcting capability. Rather, what makes RCD codes attractive is the ability to achieve high code rates coupled with the possibility of decoding these codes via a large number of decoding techniques, as per the needs of the application. Further, practical decoders for RCD codes can be implemented with lower complexity than decoders for most other codes (as an example, the SPA decoder for RCD codes has the lowest complexity among decoders for regular LDPC codes of that length). This makes them desirable candidates for implementation in high-speed applications, e.g., optical fiber communications.

Another feature that makes RCD codes particularly desirable is their tractability to combinatorial analysis, thanks to their relatively simple construction. Such analysis is helpful in obtaining bounds on RCD code performance in a variety of situations. Availability of these bounds, in turn, make RCD codes excellent candidates for the testing and validation of techniques that estimate code performance (such as DAIS and IIS). Finally, RCD codes also can serve as a starting point for constructing codes that have higher minimum distance and better performance via techniques such as serial concatenation and row-splitting [38], and they also serve as useful building blocks for constructing quantum codes that can correct for errors in the quantum domain.

C.   Recommendations For Future Research

In this section we briefly discuss the various avenues for future research that arise from this dissertation. We present the future research recommendations on a chapter by chapter basis.

Chapter III

1. The proof of the stronger results on the decoding of various classes of 3-error patterns, suggested by the analysis of data in Table III.1 and presented in Section III.F, constitutes interesting research since these results would help us ascertain the exact number of instances of decoder failures

and incorrect decoding for each class or subclass of 3-error patterns.

2. One also may perform a thorough analysis of the BFA decoding of 4-error patterns and, in particular, that of the asymptotically dominant classes, to further strengthen some of the analytical bounds.

3. The research presented in this chapter hopefully can be used to obtain bounds on the BFA decoding performance for the general class of codes based on diagonals in square arrays. In particular, it would be interesting to prove or disprove that the BFA is capable of successfully decoding all error patterns up to weight $\gamma - 1$ for a square array code constructed with γ different slope sets defining the parity-check matrix.

4) The number of decoding failures observed under BFA decoding of 3-error patterns far exceeds the minimum number of 3-error patterns that one would expect to lead to decoder failure (This number equals $\binom{6}{3}A_6$ and corresponds to all those 3-error patterns that are equidistant from the all-zeros codeword and a codeword at Hamming weight 6). This suggests the potential for improving the BFA. One possibility that immediately comes to mind is the use of a randomized rule to flip only one of all those error positions that have the maximum number of unsatisfied PCEs associated with them. Other modifications could also be attempted.

5) A combinatorial analysis of error-patterns is infeasible for codes in general, unless there is an underlying structure in the code that is simple enough to be tractable under such analysis. Another class of codes that may lend itself to combinatorial analysis of error patterns is the class of n-dimensional single parity-check product (SPCP) codes [12]. We recommend perform-

ing such a combinatorial analysis to further extend the technique proposed in this chapter and to evaluate the performance bounds for the class of n-dimensional SPCP codes.

Chapter IV

1. In general, it is not sufficient to restrict ourselves to the transmission of the all-zeros codeword in analyzing a code's decoding performance based on the e-BFA algorithm presented in Chapter IV. Hence, it would be desirable to perform a combinatorial analysis of e-error f-erasure patterns for RCD codes satisfying $2e + f = 6$ that assumes the transmission of an arbitrary codeword.

2. Modifications to the extension algorithm may be investigated with the objective of improving decoder performance. One particular suggestion involves the use of more constituent decoders (CDs), where the input to each CD is an error-pattern obtained by a unique replacement of erasures with ones and zeros.

3. A comparison between the e-BFA performance and algorithm-E performance of RCD codes at a specific high value of E_b/N_0 is very much in order, with each decoder operating at its optimal BSC/E decisioning threshold at that E_b/N_0.

4. A thorough comparison of the e-BFA algorithm and algorithm E of [65] in terms of computational complexity is certainly desirable for assessing RCD code utility. We note that the BFA (the CD for the e-BFA) requires only a single computation to be performed at each parity-check node of

the bipartite graph (an XOR operation). Algorithm E, on the other hand, requires ρ different computations (ρ is the row weight of the regular LDPC code parity-check matrix) to be performed at each parity-check node (one for each outgoing message that must be transmitted). Similarly, γ different computations are required at each variable node for algorithm E. The BFA, on the other hand, requires only a single computation at each variable node, followed by a search over all variable nodes to find the maximum. Also, the complexity of the e-BFA can be assumed to be roughly twice the complexity of the BFA (if one ignores the complexity of the decisioning that needs to be performed once both CD outputs are available).

5. It would be very useful to obtain an ASIC and/or FPGA implementation for the e-BFA decoder for certain RCD codes (such as the $\eta = 23$ RCD code) in order to evaluate RCD code performance on physical systems, e.g., optical fiber communications systems.

6. Our motivation to investigate the BSC/E arose from the desire to improve the performance of communications systems such as the optical fiber communications (OFC) system using RCD codes, without an excessive increase in computational complexity. Further, it also has been shown [62,65] that for LDPC codes a major portion of the coding gain that can be achieved in going from a BSC, with hard-decision iterative decoding, to a BPSK-AWGN channel model, with soft-decision iterative decoding, occurs in simply going to the BSC/E. The OFC channel, however, is inherently asymmetric and frequently modeled with chi-squared *pdfs* leading to a binary asymmetric channel (BAC). With this in mind, we desire to extend our

analysis to tackle the BAC and variations of the binary asymmetric channel with asymmetric erasures (BAC/AE) for RCD codes specifically, and
LDPC codes in general.

Chapter V

1. One of the immediate areas of future work, in the context of the performance curves presented in this chapter, involves the evaluation of longer
   RCD codes via DAIS or its enhancements. These evaluations will help us
   further understand the behavior of the performance of the class of RCD
   codes as a function of increasing η. These evaluations will also help us
   shed more light on the values of η for which the corresponding RCD codes
   require the least E_b/N_0 for a desired BER of say 10^{-15}.

2. The evolution of the relative contributions of decoder failures and decoding-
   to-wrong codewords to the WER at specific E_b/N_0 values, as a function
   of RCD code η, will be helpful in determining how well the SPA approximates the optimal ML decoder at those E_b/N_0 and how this approximation
   evolves as a function of η.

3. The evaluation of the RCD code performance via DAIS has provided us
   with a lot of insight into the intricacies of the DAIS technique. Certain
   limitations of the DAIS algorithm also were discovered as a result of the
   evaluations (see discussion in Section V.E). Some of the other future work
   that can be envisioned is in the area of developing and improving the DAIS
   algorithm itself, so that it is capable of evaluating longer codes in lesser
   time. These are:

(a) The evaluation of the codes presented in this chapter was performed by iteratively computing the biased *pdf* in the space of the noise random variable $\mathbf{z}$. The sum-product algorithm [41] for decoding LDPC codes requires the *log-likelihood ratios* (LLRs) [47] or the *a posteriori probability* (APP) *ratio* of each of the bits in the received word as initialization. Under a BPSK-AWGN channel model the LLRs or the APPs can be easily represented as functions of $\mathbf{z}$. We could also envision an iterative computation of the biased *pdf* in the space of LLRs or APPs, which may prove to be a more advantageous representation.

(b) The DAIS control variable for the LDPC code evaluation was chosen in an ad-hoc fashion. We would like to investigate certain other control variable definitions in order to study their relative performances as well as to determine, if possible, any optimality conditions that the control variable must satisfy.

(c) In general, the control variable need not be 1-dimensional. Admittedly, a control variable with very large dimensionality is certain to make the DAIS technique intractable. There may be certain advantages, however, in considering 2 or 3-dimensional control variables. For example, consider a 2-dimensional control variable where one of the dimensions is as defined in equation (V.1), while the other dimension corresponds to a binary variable that is an indicator for the event E. In such a case, it may be possible to eliminate the two separate simulations that need to be performed in the DAIS technique.

4. Currently there exist no analytical expressions for the variance of the error probability estimates obtained by the DAIS technique. The task of

evaluating the variance of the MMC estimator is quite complicated due to its iterative nature and the fact that the biased *pdf* is sampled using the Metropolis random walk that introduces correlation between successive samples. Some research being currently conducted on the estimator variance of MMC suggests that methods based on bootstrap techniques may provide a solution [43]. The analysis of the DAIS estimator introduces a further level of complexity since it is based on the results of two separate MMC estimates that are non-linearly combined.

Chapter VI

1. We would like to employ the IIS technique to evaluate longer codes under different hard-decision decoding algorithms in order to determine the range of applicability of the IIS technique, as well as to compare different decoding algorithms and codes.

2. The decoding analysis of lower-weight sub-patterns of an error-pattern that result in decoder error is another possible area for investigation, which would allow us to obtain more information on the decoder's ability to correct error-patterns at the lower weight.

3. The IIS technique also may be suitably modified and used to evaluate the performance of codes and decoders on the BSC/E. With such an analysis, we would be able to estimate the code performance for a range of transition probability values, without resorting to independent simulations for each choice of transition probabilities.

Chapter VII

1. We would like to obtain mathematical expressions and/or bounds for some of the values of the parameters (dimensionality and minimum-distance) for the classical codes constructed from diagonals on square $\eta \times \eta$ arrays with η prime. These results, in turn, will help us exactly determine the parameters of the quantum codes that may be constructed from such classical codes.

2. It also would be useful to obtain the minimum distances of all the classical codes possible and their duals, based on diagonals in $\eta \times \eta$ arrays, for $\eta = 11$ and 13 (the next two primes after 5 and 7). Note, however, that the task of numerically computing the minimum distance of codes becomes exceedingly time consuming very rapidly.

3. Since the classical codes based on diagonals in square arrays may be equivalently obtained from other algebraic/geometric structures such as Euclidean geometries [44] and partial BIBDs [70] (see Chapter II), it is reasonable to expect that these other classical code construction techniques also yield a class of classical codes that can be used to construct quantum codes, via the CSS technique, in a manner similar to that presented in this chapter. We recommend an investigation of these construction techniques to determine if quantum codes with superior $[n, k, d_{\min}]$ parameters are possible.

4. We need to investigate methods to further enhance the minimum distance of classical codes based on the *type-R* arrays. An example of a possible enhancement is the introduction of permutations on the columns of the *type-R* array instead of mirroring of the slope. Such enhancements may al-

together eliminate all the codewords at the lowest Hamming weight, rather than just reducing the number of codewords at the lowest Hamming weight. The process of eliminating the codeword bit-positions, thus, may be rendered unnecessary.

5. A decoder circuit for CSS quantum codes may be easily constructed once the *check matrix* [49] of the code is known. The conventional decoding scheme for quantum codes involves measuring the qubits of the quantum code with respect to measurement operators constructed based on the rows of the *check matrix*; each row of the check matrix corresponds to a specific measurement operator. The measurement outcome of the qubits of the quantum code with respect to a measurement operator is called the syndrome. The set of syndromes corresponding to the rows of the *check matrix* are then used to perform appropriate unitary transformations on the qubits of the quantum code to perform error-correction.

It also may be possible to devise low-complexity decoding schemes, however, for decoding the quantum codes resulting from our classical codes, much in the same way as such decoding schemes have been found for classical codes (such as LDPC codes). We must be mindful, of course, of the impact and limitations of quantum measurements in implementing quantum code decoders. In particular, the potentially adverse impact that repeated measurements (and, hence, interactions with the environment) can have on the state of a qubit makes it advantageous to have qubits of the quantum code participating in as few syndrome measurements as possible. The minimization of syndrome measurements per qubit may be used as a criterion to design quantum codes, and we suggest an investigation in this direction.

APPENDIX 1.

Program Listings for the IIS Technique

This appendix provides a listing of the various Matlab programs that implement the SNR-invariant importance sampling (IIS) technique described in Chapter VI. The main program is listed first followed by all secondary programs and functions.

316

bsc_invISpart1.m

| Purpose | Serves as a wrapper for the computations of the IIS technique. |

Purpose   Serves as a wrapper for the computations of the IIS technique.

Syntax   mcc -m bsc_invISpart1.m;

Description This program is written such that it allows for the creation of a stand alone C executable that implements the IIS technique. Allowing for this feature ensures a faster computation time. Upon completion of execution, the program generates a Matlab workspace file (.mat file) that contains all the statistics of the IIS simulation.

Program code

```
%This program serves as the outermost wrapper for the invariance IS
%technique
%described in Chapter VI of the Amitkumar Mahadevan's PhD dissertation.
%The BSC channel model generated by BSPK-AWGN is assumed with transition
%probability 'p'
%May 26 2004
%Amitkumar Mahadevan
%CSPL, UMBC.
%This wrappper requires the following function files in the same directory
%1. alist2Hsparse()
%2. ldpc_h2g()
%3. gen_sample_vec()
%
function [] = bsc_invISpart1();
%%%%%%%%%%%%%%%%%%%%%%%%%%%%%%%%%%%%%%%%%%%%%%%%%%%%%%%%%%%%%%%%%%%%%%%%%%%%%
%%%%%%%%%%%%%%%%%%%%%%%%%%%%%%%%%%%%%%%%%%%%%%%%%%%%%%%%%%%%%%%%%%%%%%%%%%%%%

%First all the parameters are initialized
%The structure info is used to store all the useful information

%Get the start time
clock_start = clock;

%First getting the parity-check matrix of the code
%The parity-check matrices may be stored in alist format
%Refer documentation of function alist2Hsparse() for more details
info.H_sp = alist2Hsparse('../H_matrix_289_51_sys.dat');
```

```
%systematizing the code and obtaining the G matrix
[info.H_sp_sys,info.G_sys] = ldpc_h2g(info.H_sp);

%code length and code dimensions
[info.k info.n] = size(info.G_sys);

%code rate
info.r = (info.k)/(info.n);

%generating the vector containing the number of samples for each weight
%from 0 to n
info.sample_vec = gen_sample_vec(info.n);

%and then computing the total number of samples
info.tot_samples = sum(info.sample_vec);

%Setting the maximum number of iterations for the BF decoder
info.iters_max = 20;


%%%%%%%%%%%%%%%%%%%%%%%%%%%%%%%%%%%%%%%%%%%%%%%%%%%%%%%%%%%%%%%%%%%%%%%%%%%%%%
warning off;

idx_vec = [0:1:info.n];

non_zero_sample_nos = info.sample_vec(info.sample_vec >0);
non_zero_sample_idx = idx_vec(info.sample_vec >0);

%use variable names containing at least two characters
non_zero_count = length(non_zero_sample_idx);

%initialize the word-error counters
word_error_counter = zeros(1,non_zero_count);
dwc_counter = zeros(1,non_zero_count);
failure_counter = zeros(1,non_zero_count);
MLcert_violation_counter = zeros(1,non_zero_count);

%uncoded_BER_counter is not required since this quantity is already known
%in Invariance IS and is equal to non_zero_sample_idx.*non_zero_sample_nos
%uncoded_BER_counter = zeros(1,non_zero_count);

codeword_BER_counter_manip = zeros(1,non_zero_count);
codeword_BER_counter_unmanip = zeros(1,non_zero_count);
info_BER_counter_manip = zeros(1,non_zero_count);
info_BER_counter_unmanip = zeros(1,non_zero_count);


%initialzing the average iteration counters
avg_iteration_counter = zeros(1,non_zero_count);
avg_correct_iteration_counter = zeros(1,non_zero_count);
```

```
avg_raw_success_iteration_counter = zeros(1,non_zero_count);
avg_dwc_iteration_counter = zeros(1,non_zero_count);
avg_ML_viol_iteration_counter = zeros(1,non_zero_count);

%Starting the outermost loop over all Hamming weights that are to be
%considered.

%Gathering the state of the random number generator
info.rand_state = rand('state');

%dummy definition
r_vec = [];

for ii = 1:non_zero_count

    %first check if no_of_samples equals nchoosek(code.n,ii)
    %if equal then exhasutively generate all combinations

    r = non_zero_sample_idx(ii);
    fprintf('Working on Hamming weight w=%d\n\n',r);
    if(non_zero_sample_nos(ii) < nchoosek(info.n,non_zero_sample_idx(ii)))
        fprintf('Randomly sampling error patterns since set number of ...
        error patterns less than n choose w\n\n');
    else
      if(non_zero_sample_nos(ii)==nchoosek(info.n,non_zero_sample_idx(ii)))
          fprintf('Exhaustively generating error patterns since set ...
          number of error patterns equals n choose w\n\n');
      else
          error('Number of sample vectors of weight x greater than ...
          nchoosek(n,x)');
      end
    end

      for jj= 1:non_zero_sample_nos(ii)

        %randomly generate error vectors with a fixed Hamming weight
        if(non_zero_sample_nos(ii) < ...
        nchoosek(info.n,non_zero_sample_idx(ii)))

            %generate a random permutation
            rand_vec =  randperm(info.n);

            %generate an error vector of the specific Hamming weight
            err_vec = (rand_vec <= non_zero_sample_idx(ii));

        else
            %exhaustively generate all error vectors of a specific
            %Hamming weight
        if(non_zero_sample_nos(ii)== ...
        nchoosek(info.n,non_zero_sample_idx(ii)))
```

```
  err_vec = zeros(1,info.n);
  %generating the starting error vector
  %treat the case of jj = 1 separately
  if (jj > 1)
      %Inserting the part that controls the generation of
      %the next r-combination in lexicographical order given the
      %current r-combination.
      %this part has been inserted as lines of codes
      %instead of a separate function to enhance speed

      new_r_vec = r_vec;

      if(r_vec(r)<info.n) %determining the last element of r_vec
          new_r_vec(r) = r_vec(r)+1;
      else % else applying algorithm 4.4
          for pp=(r-1):-1:1
          %finding in a_k + 1 exists in r_vec
              larger_exists = sum(r_vec(pp+1:r) == (r_vec(pp)+1));
              if(larger_exists == 0)
                  counter = 1;
                  for qq = pp:r
                      new_r_vec(qq) = r_vec(pp)+counter;
                      counter = counter+1;
                  end
                  break;
              end;
          end; %pp = (r-1):-1:1

      end; %(r_vec(r)<info.n)
      r_vec = new_r_vec;
      err_vec(r_vec) = 1;

  else%initializing the first r-combination of n
      r_vec = [1:r];
      err_vec(r_vec) = 1;
  end%(jj > 1)

else%(non_zero_sample_nos(ii)==...
  %number of error vectors to be generated greater than
  %total distinct error vectors
  error('Number of sample vectors of weight x greater than ...
  nchoosek(n,x)');
end%(non_zero_sample_nos(ii)==...
end%(non_zero_sample_nos(ii) < ...

sum_err_vec = sum(err_vec);
%the next step assumes a systematic code
info_sum_err_vec = sum(err_vec((info.n-info.k+1):info.n));
```

```
%introduce a check
if(non_zero_sample_idx(ii)~= sum_err_vec)
        error('Error: Hamming weight of generated error pattern ...
        not equal %d ', non_zero_sample_idx(ii));
end

%call the decoder. The last parameter '1' indicates that in
%case of a failure, the codeword estimate at the end of the
%last iteration will be output
[cdwd, iter_counter, success] = bitflipper(err_vec', ...
info.H_sp_sys, info.iters_max,1);
cdwd = cdwd';

sum_cdwd = sum(cdwd);
%the next step assumes a systematic code
info_sum_cdwd = sum(cdwd((info.n-info.k+1):info.n));

avg_iteration_counter(ii) = avg_iteration_counter(ii)+iter_counter;

%update the statistics
if (success == 0)%meaning decoder failure
   word_error_counter(ii) = word_error_counter(ii) + 1;
   failure_counter(ii) = failure_counter(ii) + 1;
   codeword_BER_counter_manip(ii) = ...
   codeword_BER_counter_manip(ii) + sum_cdwd;
   codeword_BER_counter_unmanip(ii) = ...
   codeword_BER_counter_unmanip(ii) + sum_err_vec;
   info_BER_counter_manip(ii) = ...
   info_BER_counter_manip(ii) + info_sum_cdwd;
   info_BER_counter_unmanip(ii) = ...
   info_BER_counter_unmanip(ii) + info_sum_err_vec;
else
   avg_raw_success_iteration_counter(ii) = ...
   avg_raw_success_iteration_counter(ii) + iter_counter;

   if (sum_cdwd~=0)%meaning DWC
       word_error_counter(ii) = word_error_counter(ii) + 1;
       dwc_counter(ii) = dwc_counter(ii) + 1;
       avg_dwc_iteration_counter(ii) = ...
       avg_dwc_iteration_counter(ii)+iter_counter;

       codeword_BER_counter_manip(ii) = ...
       codeword_BER_counter_manip(ii) + sum_cdwd;
       codeword_BER_counter_unmanip(ii) = ...
       codeword_BER_counter_unmanip(ii) + sum_cdwd;
       info_BER_counter_manip(ii) = ...
       info_BER_counter_manip(ii) + info_sum_cdwd;
       info_BER_counter_unmanip(ii) = ...
       info_BER_counter_unmanip(ii) + info_sum_cdwd;
```

```
                %in case of DWC, we look for ML certificate violation
                if( sum_err_vec <= sum(mod((err_vec+cdwd),2)) )
                %then ML certification is violated
                    MLcert_violation_counter(ii) = ...
                    MLcert_violation_counter(ii) + 1;
                    avg_ML_viol_iteration_counter(ii) = ...
                    avg_ML_viol_iteration_counter(ii) + iter_counter;
                end

                %finally resetting success to 0 to indicate that correct
                %codeword has not been reached
                success = 0;


            end%(sum_cdwd~=0)
        end%success == 0
        if(success)
            avg_correct_iteration_counter(ii) = ...
            avg_correct_iteration_counter(ii) + iter_counter ;
        end

    end%for jj= 1:non_zero_sample_nos(ii)
    %check pointing in case of premature termination of program
    save(strcat('N',num2str(info.n),'K',num2str(info.k),'temp_results'));


 end%for ii = 1:non_zero_count

clock_end = clock;
save(strcat('N',num2str(info.n),'K',num2str(info.k),'temp_results'));
```

323

bsc_invISpart2.m

Purpose	Performs postprocessing (generating the complete workspace and appropriate log file with different values of probability of error) once bsc_invISpart1.m has finished execution.
Syntax	load *.mat; bsc_invISpart2.m;
Description	This program requires the temporary workspace (.mat) file generated by bsc_invISpart1.m to be first loaded into the Matlab workspace. This program computes the various statistics associated with IIS, and performs the weighting of the raw error rates to generate the IIS error probability estimates. This program also generates a consolidated workspace (.mat) file with all the variable included, and a log file to which all the error rates and various IIS statistics are written out.

Program code

```
%This program serves as the  part 2 of the wrapper for the invariance IS
%technique described
%in Chapter VI of Amitkumar Mahadevan's PhD dissertation.
%The BSC channel model generated by BSPK-AWGN is assumed with transition
%probability 'p'
%May 26 2004
%Amitkumar Mahadevan
%CSPL, UMBC.
%This wrapper assumes that part 1 has already been executed. Part 1 performs
% the task of generating
%the error patterns at the different Hamming weights, decoding them and
%compiling the statistics
%Part 2 performs the scaling and WER and other error rate calculations for
%a range of Eb/N0 values that are required

%%%%%%%%%%%%%%%%%%%%%%%%%%%%%%%%%%%%%%%%%%%%%%%%%%%%%%%%%%%%%%%%%%%%%%%%%%%%
%%%%%%%%%%%%%%%%%%%%%%%%%%%%%%%%%%%%%%%%%%%%%%%%%%%%%%%%%%%%%%%%%%%%%%%%%%%%
%The Eb/N0 vector is the only parameter that must be initialized in part 2
%We assume that we are working with the same workspace that was present at
%the end of part 1

%BPSK AWGN Eb/N0 in dB
```

```
info.Eb_N0 = [3:0.5:15]';
%info.Eb_N0 = [3 6 8 9]';
%%%%%%%%%%%%%%%%%%%%%%%%%%%%%%%%%%%%%%%%%%%%%%%%%%%%%%%%%%%%%%%%%%%%%%%%%%

%Signal transmission is assumed at +- 1
%Computing the noise variance and std-dev
info.var = 1./(2.*(info.r).*(10.^(info.Eb_N0./10))); info.sigma =
sqrt(info.var)

%BPSK-AWGN transition probabbility
info.p = 0.5.*erfc(1./(info.sigma.*sqrt(2)));

WER_matrix = zeros(length(info.Eb_N0),non_zero_count);
dwc_rate_matrix = zeros(length(info.Eb_N0),non_zero_count);
failure_rate_matrix = zeros(length(info.Eb_N0),non_zero_count);
MLcert_violation_rate_matrix =
zeros(length(info.Eb_N0),non_zero_count);
codeword_BER_manip_matrix =
zeros(length(info.Eb_N0),non_zero_count);
codeword_BER_unmanip_matrix =
zeros(length(info.Eb_N0),non_zero_count); info_BER_manip_matrix =
zeros(length(info.Eb_N0),non_zero_count); info_BER_unmanip_matrix
= zeros(length(info.Eb_N0),non_zero_count); weight_matrix =
zeros(length(info.Eb_N0),non_zero_count);

    %Computation of WER based on equation VI.13 of PhD proposal by
    %Amitkumar Mahadevan

    for kk = 1:non_zero_count

    weight_matrix(:,kk) = ((info.p).^non_zero_sample_idx(kk))...
    .*((1-info.p).^(info.n - non_zero_sample_idx(kk)))...
                .*nchoosek(info.n, non_zero_sample_idx(kk));

    WER_matrix(:,kk) = (word_error_counter(kk)./...
    non_zero_sample_nos(kk)).*weight_matrix(:,kk);
    failure_rate_matrix(:,kk) = (failure_counter(kk)./...
    non_zero_sample_nos(kk)).*weight_matrix(:,kk);
    dwc_rate_matrix(:,kk) = (dwc_counter(kk)./...
    non_zero_sample_nos(kk)).*weight_matrix(:,kk);
    MLcert_violation_rate_matrix(:,kk) =  ...
    (MLcert_violation_counter(kk)./non_zero_sample_nos(kk))...
    .*weight_matrix(:,kk);

    codeword_BER_manip_matrix(:,kk) = (codeword_BER_counter_manip(kk)...
    ./(non_zero_sample_nos(kk).*info.n)).*weight_matrix(:,kk);
    codeword_BER_unmanip_matrix(:,kk) = ...
    (codeword_BER_counter_unmanip(kk)...
    ./(non_zero_sample_nos(kk).*info.n)).*weight_matrix(:,kk);
    info_BER_manip_matrix(:,kk) = (info_BER_counter_manip(kk)...
```

```
          ./(non_zero_sample_nos(kk).*info.k)).*weight_matrix(:,kk);
      info_BER_unmanip_matrix(:,kk) = (info_BER_counter_unmanip(kk)...
          ./(non_zero_sample_nos(kk).*info.k)).*weight_matrix(:,kk);

   end

   WER = sum(WER_matrix,2);
   failure_rate = sum(failure_rate_matrix,2);
   dwc_rate = sum(dwc_rate_matrix,2);
   MLcert_violation_rate = sum(MLcert_violation_rate_matrix,2);

   codeword_BER_manip = sum(codeword_BER_manip_matrix,2);
   codeword_BER_unmanip = sum(codeword_BER_unmanip_matrix,2);
   info_BER_manip = sum(info_BER_manip_matrix,2);
   info_BER_unmanip = sum(info_BER_unmanip_matrix,2);

   %Computing the bias and variance of the estimator
   temp1_var_matrix = weight_matrix.^2.*WER_matrix;
   temp2_var_matrix = zeros(size(temp1_var_matrix));
   for (ww=1:non_zero_count)
      temp2_var_matrix(:,ww) = temp1_var_matrix(:,ww)...
      ./non_zero_sample_nos(ww);
   end

   variance_vector = sum(temp2_var_matrix,2);

    % Data output to file and screen
fstring_summary =
strcat('N',num2str(info.n),'K',num2str(info.k),'results',
num2str(clock_end(2)),
num2str(clock_end(3)), num2str(clock_end(1)),'.dat');

fid_summary = fopen(fstring_summary,'w');

%setting the printing format
format long e;

%display statements
fprintf('Start Time = %2d/%2d/%4d (MM/DD/YYYY), ...
\t %2d:%2d:%5.2f (hrs:min:sec)\n',...
clock_start(2), clock_start(3), clock_start(1),clock_start(4),...
clock_start(5), clock_start(6));
fprintf('End Time = %2d/%2d/%4d (MM/DD/YYYY), ...
\t %2d:%2d:%5.2f (hrs:min:sec)\n\n',...
clock_end(2), clock_end(3), clock_end(1), clock_end(4),...
clock_end(5), clock_end(6));
fprintf(fid_summary,'Start Time = %2d/%2d/%4d (MM/DD/YYYY), ...
\t %2d:%2d:%5.2f (hrs:min:sec)\n',...
clock_start(2), clock_start(3), clock_start(1), clock_start(4),...
clock_start(5), clock_start(6));
```

```
fprintf(fid_summary,'End Time = %2d/%2d/%4d (MM/DD/YYYY), ...
\t %2d:%2d:%5.2f (hrs:min:sec)\n\n',...
clock_end(2), clock_end(3), clock_end(1), clock_end(4),...
clock_end(5), clock_end(6));

fprintf('Invariance IS results for the n = %d, k = %d code ...
under bit-flipping decoding for the BSC\n\n',info.n,info.k);
fprintf(fid_summary,'Invariance IS results for the n = %d, k = %d code ...
under bit-flipping decoding for the BSC\n\n',info.n,info.k);

fprintf('Total codewords transmitted = %d\n',info.tot_samples);
fprintf('Total code bits transmitted = %d\n',(info.tot_samples.*info.n));
fprintf('Total information bits transmitted = %d\n\n',...
(info.tot_samples.*info.k));
fprintf(fid_summary,'Total codewords transmitted = %d\n',info.tot_samples);
fprintf(fid_summary,'Total code bits transmitted = %d\n',...
(info.tot_samples.*info.n));
fprintf(fid_summary,'Total information bits transmitted = %d\n\n',...
(info.tot_samples.*info.k));

fprintf('Hamming weights with non-zero codewords transmitted
are\n');
fprintf('%d ',non_zero_sample_idx);
fprintf('\nCorresponding number of codewords trasnmitted are \n');
fprintf('%d ',non_zero_sample_nos);
fprintf('\nCodewords in error at each Hamming weight considered
are\n');
fprintf('%d ',word_error_counter);
fprintf('\nInstances of decoder failure at each Hamming weight
considered are\n');
fprintf('%d ',failure_counter);
fprintf('\nInstances of decoding to wrong codeword (DWC) at each
Hamming weight considered are\n');
fprintf('%d ',dwc_counter);
fprintf('\nInstances of ML certificate violation at each Hamming
weight considered are\n');
fprintf('%d ',MLcert_violation_counter);
fprintf('\nUnmanipulated codeword bits in error at each Hamming
weight considered are\n');
fprintf('%d ',codeword_BER_counter_unmanip);
fprintf('\nManipulated codeword bits in error at each Hamming
weight considered are\n');
fprintf('%d ',codeword_BER_counter_manip);
fprintf('\nUnmanipulated info bits in error at each Hamming weight
considered are\n');
fprintf('%d ',info_BER_counter_unmanip);
fprintf('\nManipulated info bits in error at each Hamming weight
considered are\n');
fprintf('%d ',info_BER_counter_manip);
fprintf('\n\nAverage iterations to decode at each Hamming weight
```

```
considered are\n');
fprintf('%f ',avg_iteration_counter./(non_zero_sample_nos));

fprintf(fid_summary,'Hamming weights with non-zero codewords
transmitted are\n');
fprintf(fid_summary,'%d ',non_zero_sample_idx);
fprintf(fid_summary,'\nCorresponding number of codewords
trasnmitted are \n');
fprintf(fid_summary,'%d ',non_zero_sample_nos);
fprintf(fid_summary,'\nCodewords in error at each Hamming weight
considered are\n');
fprintf(fid_summary,'%d ',word_error_counter);
fprintf(fid_summary,'\nInstances of decoder failure at each
Hamming weight considered are\n');
fprintf(fid_summary,'%d ',failure_counter);
fprintf(fid_summary,'\nInstances of decoding to wrong codeword
(DWC) at each Hamming weight considered are\n');
fprintf(fid_summary,'%d ',dwc_counter);
fprintf(fid_summary,'\nInstances of ML certificate violation at
each Hamming weight considered are\n');
fprintf(fid_summary,'%d ',MLcert_violation_counter);
fprintf(fid_summary,'\nUnmanipulated codeword bits in error at
each Hamming weight considered are\n');
fprintf(fid_summary,'%d ',codeword_BER_counter_unmanip);
fprintf(fid_summary,'\nManipulated codeword bits in error at each
Hamming weight considered are\n');
fprintf(fid_summary,'%d ',codeword_BER_counter_manip);
fprintf(fid_summary,'\nUnmanipulated info bits in error at each
Hamming weight considered are\n');
fprintf(fid_summary,'%d ',info_BER_counter_unmanip);
fprintf(fid_summary,'\nManipulated info bits in error at each
Hamming weight considered are\n');
fprintf(fid_summary,'%d ',info_BER_counter_manip);
fprintf(fid_summary,'\n\nAverage iterations to decode at each
Hamming weight considered are\n');
fprintf(fid_summary,'%f ',avg_iteration_counter./(non_zero_sample_nos));

fprintf('\nAverage iterations to decode when correct, at each
Hamming weight considered are\n');
fprintf('%f ',avg_correct_iteration_counter./
(non_zero_sample_nos-failure_counter-dwc_counter));
fprintf('\nAverage iterations to decode when DWC, at each Hamming
weight considered are\n');
fprintf('%f ',avg_dwc_iteration_counter./dwc_counter);
fprintf('\nAverage iterations to decode when raw success, at each
Hamming weight considered are\n');
fprintf('%f ',avg_raw_success_iteration_counter./
(non_zero_sample_nos-failure_counter));
fprintf('\nAverage iterations to decode when ML certificate
violated, at each Hamming weight considered are\n');
```

```
fprintf('%f ',avg_ML_viol_iteration_counter./MLcert_violation_counter);
fprintf('\n\n');


fprintf(fid_summary,'\nAverage iterations to decode when correct,
at each Hamming weight considered are\n');
fprintf(fid_summary,'%f ',avg_correct_iteration_counter./
(non_zero_sample_nos-failure_counter-dwc_counter));
fprintf(fid_summary,'\nAverage iterations to decode when DWC, at
each Hamming weight considered are\n');
fprintf(fid_summary,'%f ',avg_dwc_iteration_counter./dwc_counter);
fprintf(fid_summary,'\nAverage iterations to decode when raw
success, at each Hamming weight considered are\n');
fprintf(fid_summary,'%f ',avg_raw_success_iteration_counter./
(non_zero_sample_nos-failure_counter));
fprintf(fid_summary,'\nAverage iterations to decode when ML
certificate violated, at each Hamming weight considered are\n');
fprintf(fid_summary,'%f ',avg_ML_viol_iteration_counter
./MLcert_violation_counter);
fprintf(fid_summary,'\n\n');

for kk = 1:length(info.Eb_NO)
fprintf('Eb_NO = %f dB, AWGN variance = %f, BSC transition probability,
 p = %e\n',info.Eb_NO(kk),info.var(kk),info.p(kk));
fprintf(fid_summary,'Eb_NO = %f dB, AWGN variance = %f, BSC transition
 probability,p = %e\n',info.Eb_NO(kk),info.var(kk),info.p(kk));

%fprintf('Uncoded BER = %f\n', uncoded_BER);
fprintf('WER = %e, codeword failure rate = %e, DWC rate = %e \n',
WER(kk), failure_rate(kk),dwc_rate(kk));
fprintf('ML certificate violation rate = %e\n',MLcert_violation_rate(kk));
fprintf('Manipulated info BER = %e, Unmanipulated info BER = %e\n',
info_BER_manip(kk),info_BER_unmanip(kk));
fprintf('Manipulated codeword BER = %e, Unmanipulated codeword BER = %e\n',
codeword_BER_manip(kk),codeword_BER_unmanip(kk));
fprintf('\n');
%fprintf('Uncoded BER = %f\n', uncoded_BER);
fprintf(fid_summary,'WER = %e, codeword failure rate = %e, DWC rate
= %e \n',WER(kk), failure_rate(kk),dwc_rate(kk));
fprintf(fid_summary,'ML certificate violation rate = %e\n\n',
MLcert_violation_rate(kk));
fprintf(fid_summary,'Manipulated info BER = %e, Unmanipulated info BER
= %e\n',info_BER_manip(kk),info_BER_unmanip(kk));
fprintf(fid_summary,'Manipulated codeword BER = %e,
Unmanipulated codeword BER = %e\n', codeword_BER_manip(kk),
codeword_BER_unmanip(kk));
fprintf(fid_summary,'\n');

end; warning off;

fclose(fid_summary);
```

```
save(strcat('N',num2str(info.n),'K',num2str(info.k),'results',
num2str(clock_end(2)), num2str(clock_end(3)),
num2str(clock_end(1))));

figure(2) for kk = 1:length(info.Eb_N0)
semilogy(non_zero_sample_idx,WER_matrix(kk,:),'b-o','linewidth',2);
hold on;
txtstr = strcat(num2str(info.Eb_N0(kk)),' dB');
hold on;

end grid on
semilogy(non_zero_sample_idx,word_error_counter./non_zero_sample_nos);
figure(3); semilogy(info.Eb_N0,WER,'k-*','linewidth',2) hold on;
semilogy(info.Eb_N0,info_BER_manip,'b:*','linewidth',2) hold on;
semilogy(info.Eb_N0,info_BER_unmanip,'b--*','linewidth',2) hold
on; semilogy(info.Eb_N0,dwc_rate,'r-o','linewidth',2) hold on;
semilogy(info.Eb_N0,MLcert_violation_rate,'k+','linewidth',2)
legend('WER','information BER (manipulated)','information BER
(unmanipulated)','DWC rate','ML certificate violation rate'); grid
on; ylabel('Error probability'); xlabel('\it{E_{b}}/\it{N}_{0}');

%BDD WER computation based on correction of all error patterns up to weight
%defined by the minimum distance of the code
% non_zero_sample_idx must be set to be greater than the bdd capacity

BDD_WER = sum(weight_matrix(:,(non_zero_sample_idx>2)),2);
%Plots for allerton submission
%First the weighted error probability vs. Hamming weight plot
figure(122)
raw_WER = word_error_counter./non_zero_sample_nos;

semilogy(non_zero_sample_idx,raw_WER,'ko','linewidth',2,'markersize',6);
hold on; txtstr = 'Raw word-error probability';

semilogy(non_zero_sample_idx,WER_matrix(1,:),'k-s','linewidth',2,
'markersize',6); hold on; txtstr = strcat(num2str(info.Eb_N0(1)),'
dB');

legh = legend('Raw word-error probability','Weighted word-error
probability'); set(legh,'fontsize',18,'fontname','times');

for kk = 2:length(info.Eb_N0)
semilogy(non_zero_sample_idx,WER_matrix(kk,:),'k-s','linewidth',2,
'markersize',6);
hold on;

txtstr = strcat(num2str(info.Eb_N0(kk)),' dB'); hold on; end

grid on;
```

```
set(gca,'TickLength', [0.02 0.02]) set(gca,'FontSize', 27)
set(gca,'FontName', 'Times') set(gca,'LineWidth', 2) ylabel('Error
Probability'); xlabel('Hamming weight'); ylim([1e-50 1]); set(gca,
'ytick', [1e-50 1e-40 1e-30 1e-20 1e-10 1]);

figure(123)
semilogy(info.Eb_N0,info.p,'k.','linewidth',2,'markersize',12);
hold on;
semilogy(info.Eb_N0,WER,'k-*','linewidth',2,'markersize',12) hold
on;
semilogy(info.Eb_N0,info_BER_unmanip,'k:o','linewidth',2,'markersize',12)
hold on;
semilogy(info.Eb_N0,BDD_WER,'k--+','linewidth',2,'markersize',12)
grid on;

grid on; legh = legend('BSC transition probability, {\it p}','BFA
WER by IIS', 'BFA BER by IIS', 'Theoretical BDD WER');
set(legh,'fontsize',18,'fontname','times');

set(gca,'TickLength', [0.02 0.02]) set(gca,'FontSize', 27)
set(gca,'FontName', 'Times') set(gca,'LineWidth', 2) ylabel('BER
or WER'); xlabel('E_{b}/N_{0} dB'); xlim([2 14]); ylim([1e-50 1]);
set(gca, 'ytick', [1e-50 1e-40 1e-30 1e-20 1e-10 1]);
```

bsc_invISplot.m

Purpose   Generates plots related to the IIS simulation. Also generates the code performance curve.

Syntax   `load *.mat; bsc_invISplot.m;`

Description This program requires the consolidated workspace (.mat) file generated by `bsc_invISpart2.m` to be first loaded into the Matlab workspace. This is essentially a plotting program, which generates the code performance curve amongst other plots.

Program code

```
%This program serves as the  plotting function of the wrapper for the
%invariance IS technique described
%in Chapter VI of the Amitkumar Mahadevan's PhD dissertation.
%The BSC channel model generated by BSPK-AWGN is assumed with
%transition probability 'p'
%May 26 2004
%Amitkumar Mahadevan
%CSPL, UMBC.
%This wrapper assumes that part 1 and part 2 have already been executed.
%Part 1 performs the task of generating
%the error patterns at the different Hamming weights, decoding them and
%compiling the statistics
%Part 2 performs the scaling and WER and other error rate calculations for
%a range of Eb/N0 values that are required
%this wrapper (which may be considered as part 3) generates the relevant
%plots and also includes the results of any standard Monte Carlo
%simulations performed for comparison
%Two basic types of plots are generated
%1. Error Probability vs. Hamming weight
%2. BER and WER vs. Eb/N0 including std. Monte Carlo results
%The confidence intervals presented in the standard MC results are based on
%the Gaussian Approximation, which is valid since we collect a large number
%of error events (usually 100)
%Make sure to load the results of part 1 and part 2 before proceeding with
%the part for plotting

%%%%%%%%%%%%%%%%%%%%%%%%%%%%%%%%%%%%%%%%%%%%%%%%%%%%%%%%%%%%%%%%%%%%%%%%%%%%%%
%%%%%%%%%%%%%%%%%%%%%%%%%%%%%%%%%%%%%%%%%%%%%%%%%%%%%%%%%%%%%%%%%%%%%%%%%%%%%%
%The Eb/N0 vector is one parameter that must be initialized in the
%'plot' part
```

```
%We assume that we are working with the same workspace that was present at
%the end of part 1

%BPSK AWGN Eb/NO in dB
%info.Eb_NO = [3:0.5:15]';
info.Eb_NO = [3 5 11 14]';

%Flag to indicate if standard MC data is available for plotting
%If true (=1) then program will attempt to load
%mat file with the name and directory path specified
std_MC_available = 1; std_MC_str = 'N256K81results/std_MC_data';

%Confidence interval specifications
%confidence level as a fraction
conf_level = 0.99;
%computing d_alpha based on eq. 5.6.11 in Jecruchim Balaban and Shanmugham
% (JBS): simulations of Communications Systems
%also refer green note-book Oct30 2003 equation I
d_alpha = sqrt(2)*erfcinv(1-conf_level);

%%%%%%%%%%%%%%%%%%%%%%%%%%%%%%%%%%%%%%%%%%%%%%%%%%%%%%%%%%%%%%%%%%%%%%%%%%%%%
%Signal transmission is assumed at +- 1
%Computing the noise variance and std-dev
info.var = 1./(2.*(info.r).*(10.^(info.Eb_NO./10))); info.sigma =
sqrt(info.var)

%BPSK-AWGN transition probabbility
info.p = 0.5.*erfc(1./(info.sigma.*sqrt(2)));

WER_matrix = zeros(length(info.Eb_NO),non_zero_count);
dwc_rate_matrix = zeros(length(info.Eb_NO),non_zero_count);
failure_rate_matrix = zeros(length(info.Eb_NO),non_zero_count);
MLcert_violation_rate_matrix =
zeros(length(info.Eb_NO),non_zero_count);

codeword_BER_manip_matrix =
zeros(length(info.Eb_NO),non_zero_count);
codeword_BER_unmanip_matrix =
zeros(length(info.Eb_NO),non_zero_count); info_BER_manip_matrix =
zeros(length(info.Eb_NO),non_zero_count); info_BER_unmanip_matrix
= zeros(length(info.Eb_NO),non_zero_count); weight_matrix =
zeros(length(info.Eb_NO),non_zero_count);

    %Computation of WER based on equation VI.13 of PhD proposal by
    %Amitkumar Mahadevan

    for kk = 1:non_zero_count

    weight_matrix(:,kk) =
    ((info.p).^non_zero_sample_idx(kk)).*((1-info.p)...
```

```
                        .^(info.n - non_zero_sample_idx(kk)))...
                        .*nchoosek(info.n, non_zero_sample_idx(kk));
    WER_matrix(:,kk) = (word_error_counter(kk)./non_zero_sample_nos(kk))...
    .*weight_matrix(:,kk);
    failure_rate_matrix(:,kk)=(failure_counter(kk)...
    ./non_zero_sample_nos(kk)).*weight_matrix(:,kk);
    dwc_rate_matrix(:,kk) = (dwc_counter(kk)./non_zero_sample_nos(kk))...
    .*weight_matrix(:,kk);
    MLcert_violation_rate_matrix(:,kk) = (MLcert_violation_counter(kk)...
    ./non_zero_sample_nos(kk)).*weight_matrix(:,kk);

    codeword_BER_manip_matrix(:,kk) = (codeword_BER_counter_manip(kk)...
    ./(non_zero_sample_nos(kk).*info.n)).*weight_matrix(:,kk);
    codeword_BER_unmanip_matrix(:,kk) =(codeword_BER_counter_unmanip(kk)...
    ./(non_zero_sample_nos(kk).*info.n)).*weight_matrix(:,kk);
    info_BER_manip_matrix(:,kk) = (info_BER_counter_manip(kk)...
    ./(non_zero_sample_nos(kk).*info.k)).*weight_matrix(:,kk);
    info_BER_unmanip_matrix(:,kk) = (info_BER_counter_unmanip(kk)...
    ./(non_zero_sample_nos(kk).*info.k)).*weight_matrix(:,kk);

    end

    WER = sum(WER_matrix,2);
    failure_rate = sum(failure_rate_matrix,2);
    dwc_rate = sum(dwc_rate_matrix,2);
    MLcert_violation_rate = sum(MLcert_violation_rate_matrix,2);

    codeword_BER_manip = sum(codeword_BER_manip_matrix,2);
    codeword_BER_unmanip = sum(codeword_BER_unmanip_matrix,2);
    info_BER_manip = sum(info_BER_manip_matrix,2);
    info_BER_unmanip = sum(info_BER_unmanip_matrix,2);

    % The 'plot' part does not do any data output to file and screen
    %Neither is any data stored to the workspace!
%BDD WER computation based on correction of all error patterns up to weight
%7 for the SPCP code and weight 8 for the EG code
BDD_WER = sum(weight_matrix(:,(non_zero_sample_idx>=8)),2);

%Below values are hard-coded for the 4-D SPCP code and correspond
%to 3,3.5,4, and 4.5 dB
%Comment or delete this line for other codes
BDD_WER(1) = 1; BDD_WER(2) = 1; BDD_WER(3); 1; BDD_WER(4) = 1;

%Laoding Standard MC data
%rememeber to run std_MC_data_gen.m to first create the std_MC_data.mat
%file and then load it here if(std_MC_available)
    load(std_MC_str);
    std_MC_WER = std_MC_word_error_count./std_MC_words_xmitted;
    std_MC_unmanip_infoBER = std_MC_unmanip_infobit_error_count...
    ./std_MC_infobits_xmitted;
```

```
%We need to compute the confidence intervals based on the Gaussian
%approximation
eta_vec = std_MC_word_error_count;
temp_vec = sqrt(((4.*eta_vec)./(d_alpha^2))+1);
upper_interim_vec = 1 + (((d_alpha^2)./(2.*eta_vec)).*(1+temp_vec));
lower_interim_vec = 1 + (((d_alpha^2)./(2.*eta_vec)).*(1-temp_vec));

%upper limit of confidence interval on ch_ber
% y_plus_ch_ber = Our_standard_MC_in_Cpp_channel_Ber.*upper_interim_vec;
% y_minus_ch_ber = Our_standard_MC_in_Cpp_channel_Ber.*lower_interim_vec;
% %upper limit of confidence interval on info_ber
% y_plus_info_ber = Our_standard_MC_in_Cpp_info_Ber.*upper_interim_vec;
% y_minus_info_ber = Our_standard_MC_in_Cpp_info_Ber.*lower_interim_vec;

y_plus_Wer = std_MC_WER.*upper_interim_vec; y_minus_Wer =
std_MC_WER.*lower_interim_vec;

end

%Below we are interested in generating plots for conference or journal
%papers
%%First the weighted error probability vs. Hamming weight plot
figure(122); raw_WER = word_error_counter./non_zero_sample_nos;

semilogy(non_zero_sample_idx,raw_WER,'ko','linewidth',2,'markersize',6);
hold on;
%txtstr = 'Raw word-error probability';

semilogy(non_zero_sample_idx,WER_matrix(1,:),'k-s','linewidth',2,
'markersize',6);
hold on;
txtstr = strcat(num2str(info.Eb_N0(1)),' dB');
text(non_zero_sample_idx(non_zero_count-1),
WER_matrix(1,non_zero_count-1),txtstr,'backgroundcolor','w',
'fontname','times','fontsize',16);

legh = legend('Raw WER','Weighted WER');
set(legh,'fontsize',18,'fontname','times');

for kk = 2:length(info.Eb_N0)
semilogy(non_zero_sample_idx,WER_matrix(kk,:),'k-s','linewidth',2,
'markersize',6);
hold on;

txtstr = strcat(num2str(info.Eb_N0(kk)),' dB');
text(non_zero_sample_idx(non_zero_count-1),
WER_matrix(kk,non_zero_count-1),txtstr,'backgroundcolor','w',
'fontname','times','fontsize',16);
hold on; end
```

```
grid on;

set(gca,'TickLength', [0.02 0.02]) set(gca,'FontSize', 27)
set(gca,'FontName', 'Times') set(gca,'LineWidth', 2) ylabel('Error
Probability at {\it w}'); xlabel('Hamming weight, {\it w}');
ylim([1e-25 1]); set(gca, 'ytick', [1e-25 1e-20 1e-15 1e-10 1e-5
1]);

figure(123)
semilogy(info.Eb_N0,info.p,'k.','linewidth',2,'markersize',12);
hold on;
semilogy(info.Eb_N0,WER,'k-*','linewidth',2,'markersize',12) hold
on;
semilogy(info.Eb_N0,info_BER_unmanip,'k:o','linewidth',2,'markersize',12)
hold on;
semilogy(info.Eb_N0,BDD_WER,'k--+','linewidth',2,'markersize',12);
hold on; grid on;

if(std_MC_available);
    semilogy(std_MC_Eb_N0,std_MC_WER,'ks','linewidth',2,'markersize',12);
    legh = legend('BSC transition probability, {\it p}','BFD WER by IIS',
    'BFD BER by IIS', 'Theoretical BDD WER', 'BFD WER by standard MC');
    set(legh,'fontsize',18,'fontname','times');

    %plotting the confidence interval
    for iii= 1:length(eta_vec)
        line([std_MC_Eb_N0(iii), std_MC_Eb_N0(iii)], [y_minus_Wer(iii),...
        y_plus_Wer(iii)], 'linewidth',2);
    end
else
    legh = legend('BSC transition probability, {\it p}','BFD WER by IIS',
    'BFD BER by IIS', 'Theoretical MLD WER');
    set(legh,'fontsize',18,'fontname','times');
end

set(gca,'TickLength', [0.02 0.02]) set(gca,'FontSize', 27)
set(gca,'FontName', 'Times') set(gca,'LineWidth', 2) ylabel('BER
or WER'); xlabel('E_{b}/N_{0} dB'); xlim([2 15]); ylim([1e-30 1]);
set(gca, 'ytick', [1e-30 1e-25 1e-20 1e-15 1e-10 1e-05 1]);
```

alist2Hsparse.m

Purpose	Reads the LDPC code parity-check matrix stored in a text file in the *alist* format [48] and converts it to the Matlab *sparse matrix* format.
Syntax	H_sparse = alist2Hsparse(alist_filename);
Description	This function is used by bsc_invISpart1.m.
	H_sparse: Code parity-check matrix in Matlab *sparse matrix* format.
	alist_filename: The parity-check matrix stored as a text file in the *alist* format.

Program code

```
function H_sparse = alist2Hsparse( alist_filename);
%Amitkumar Mahadevan, Sep 2003
%copyright 2003
%communications and signal processing laboratory
%University of Maryland Baltimore County
%  version 1: 03 Oct 2003
%  Purpose: Convert the parity-check matrix from the
%            alist format of D. J. C. MAcKay to a parity
%            check matrix in systematic form and Matlab sparse format.
%  Inputs : 1. alist_filename : A string that represents the name of the
%            file
%            in which the H matrix is stored in the alist format
%  Outputs: 1. H_sparse: the parity-check matrix (H matrix) of a regular
%            LDPC code in sparse format.
%            Note that the program checks to see if the provided H matrix
%            corresponds to a row and
%            column regular LDPC code. An error message is given and the
%            program is terminated if this condition
%            is not satisfied.

%The alist format
% Let n = code length, k = information length, m = no. of rows of the H
%matrix.
% Let gamma = coulmn weight of all columns of the H matrix.
%(i.e., the number of ones in each column of the H-matrix)
% Let rho = row weight of all rows of the H matrix.
% The alist file displays the following

% /beginning of file
```

```
% 1st line:    n m k
% 2nd line:    gamma rho
% 3rd line:    row indices in ascending order corresponding to the
%location of 1s in the first column
% 4th line:    row indices in ascending order corresponding to the
%location of 1s in the second column
%         .
%         .
%         .
% (n+2)th line: row indices in ascending order corresponding to the
%location of 1s in the nth column
% (n+3)th line: column indices in ascending order corresponding to
%location of 1s in the first row
% (n+4)th line: column indices in ascending order corresponding to
%location of 1s in the second row
%         .
%         .
%         .
% (n+m+2)th line: column indices in ascending order corresponding to
%location of 1s in the mth row.
% /end of file
%NOTE THAT INDEXING BEGINS AT 1

%An example of an alist file corresponding to n = 4 (no. of columns),
%m = 4 (no. of rows), k = 1 code with gamma = rho = 2
%and with a parity check matrix given by
%  [1 1 0 0
%   0 0 1 1
%   1 0 1 0
%   0 1 0 1];
% In general, one needs to compute the rank of the H-matrix in order
%to determine k
% This can also be done by converting the H-matrix to a systematic G
%matrix. Then, k equals
% the number of rows of the G matrix

%Begin alist file
% 4 4 1
% 2 2
% 1 3
% 1 4
% 2 3
% 2 4
% 1 2
% 3 4
% 1 3
% 2 4
%End alist file
%%%%%%%%%%%%%%%%%%%%%%%%%%%%%%%%%%%%%%%%%%%%%%%%%%%%%%%%%%%%%%%%%%%%%%%%%%
```

```
%opening file
fid = fopen(alist_filename,'r');
n=0;m=0;k=0;col_wt_min=0;row_wt_min=0;
%reading the code parameters n, m, k and gamma and rho
[inp_vector1,count]=fscanf(fid,'%d',3);
if(count ~=3)
    error('First line of a list file must contain 3 inputs');
end n = inp_vector1(1) m= inp_vector1(2) k = inp_vector1(3)

[inp_vector2, count] = fscanf(fid,'%d',2);
if(count ~=2)
    error('Second line of a list file must contain 2 inputs');
end
col_wt_min = inp_vector2(1) row_wt_min = inp_vector2(2)

%the parity check matrix has m rows and n cols
%two matrices are initialized to read the column indices
%in one, and the row indices in the other
%The alist format is redundant and so eventually H_full_1 must equal
%to H_full_2
H_full_1 = zeros(m,n); H_full_2 = zeros(m,n);

%now reading the row indices corresponding to the ones
%in each of the n columns
for i= 1:n
    read_row_indices = [];
    [read_row_indices count]=fscanf(fid,'%d',col_wt_min);
    %fscanf(fid,'\n');
    if(count~=col_wt_min)
        error('inconsistency in number of entries read for each
        column and the column weight\n');
    end
    %writing the first H matrix
    for j = 1:col_wt_min
        H_full_1(read_row_indices(j),i) = 1;
    end
end

for i= 1:m
    read_column_indices = [];
    [read_column_indices count] = fscanf(fid,'%d',row_wt_min);
    %fscanf(fid,'\n');
    if(length(read_column_indices)~=row_wt_min)
        error('inconsistency in number of entries read for each
        row and the row weight\n');
    end
    %writing the first H matrix
    for j = 1:row_wt_min
        H_full_2(i,read_column_indices(j)) = 1;
    end
```

```
end

fclose(fid);
if (sum(sum(H_full_1 == H_full_2)) ~= m*n)
    error('H_full1 and H_full2 are different');
end
H_sparse = sparse(H_full_1);
%end function
```

gen_sample_vec.m

Purpose Initializes the number of received words (error-patterns) that must be generated and decoded at each Hamming weight.

Syntax sample_number_vec = gen_sample_vec(n);

Description This function is used by bsc_invISpart1.m.

sample_number_vec: A vector of length $n+1$ that stores the number of error-patterns that must be decoded at each Hamming weight.
n: The length of the code whose performance is being evaluated via the IIS technique.

Program code

```
%This function generates a vector of the number of error vectors that must
%be generated for Invariance IS for each codeword Hamming weight
%The input is the length 'n' of the code
%The output is a vector of length 'n+1'
%See Chapter VI of the Amitkumar Mahadevan's PhD dissertation. Also see the
%documentation of the program bsc_invIS.m
%The number of samples for any Hamming weight may be equal to 0
%%May 26 2004 Amitkumar Mahadevan CSPL, UMBC.
function sample_number_vec = gen_sample_vec(n);

warning off;%turning warnings off
%initializing the number of samples vector
sample_number_vec = zeros(1,n+1);
for ii = 4:7
    sample_number_vec(ii) = min(nchoosek(n,ii-1),1e7);
end
for ii = 8:15
    sample_number_vec(ii) = min(nchoosek(n,ii-1),1e6);
end
for ii = 16:21
    sample_number_vec(ii) = min(nchoosek(n,ii-1),1e5);
end
for ii = 21:50
    sample_number_vec(ii) = min(nchoosek(n,ii-1),1e4);
end
return;
```

bitflipper.m

Purpose   Implements the bit-flipping algorithm for decoding or-
                  thogonal column-regular LDPC codes [38].

Syntax   [cdwd, iter_counter, success] =
                  bitflipper(rcvd_wd, H_sp, max_iters,
                  pass_flipped)

Description   This function is used by bsc_invISpart1.m.

                  cdwd: The codeword estimate output of the bit-flipping
                  algorithm
                  iter_counter: The number of iterations taken to de-
                  code the received word.
                  success: A boolean variable which is true if the bit-
                  flipping algorithm converges to a valid codeword
                  rcvd_word: The received word that is input to the de-
                  coder.
                  H_sp: The parity-check matrix of the code being evalu-
                  ated in Matlab *sparse matrix* format.
                  max_iters: The maximum number of iterations permit-
                  ted for the bit-flipping algorithm.
                  pass_flipped: Determines what vector is output by the
                  bit-flipping algorithm in the event of a decoder failure;
                  if set to true, the modified word at the end of the last
                  iteration is output by the function; if set to false, the
                  input received word is passed out as the decoder output.

Program code

```
function [cdwd, iter_counter, success] = bitflipper(rcvd_wd, H_sp,
max_iters, pass_flipped)

%This program simulates the Bit Flipping algorithm as proposed
%by Gallager
%rcvd_wd is a column vector that represents the received bit stream
%H_sp is the parity-check matrix of the code in which the entries are
%either ones %or zeros. H_sp is in sparse format
%max_iters is the maximum number of iterations permitted
%if pass_flipped is '1', we propagate the manipulated received word in case
%the decoder does not successfully decode the received word.
%if pass_flipped is 0, we propagate the orignal received word in case of a
```

```
%decoder failure
%cdwd is the decoder output
%iter_counter is the number of iterations required by the decoder
%success is a flag that determines if the decoder succeeded in arriving at
%a valid
%codeword or not

%converting to full
H = full(H_sp);

% the length of the codeword i.e. the number of columns of H
N = size(H,2);

%storing the original received word
rcvd_wd_orig = rcvd_wd;

% we multiply the received word with the H matrix.
% and take the resultant modulo 2
eq_sol = mod((H*rcvd_wd),2);

%setting the iteration counter
iter_counter = 0;

% checking to see if any of the parity checks are unsatisfied.
while ( sum(eq_sol) ~= 0 & (iter_counter < max_iters))

    iter_counter = iter_counter + 1;
    % finding the number of unsatisfied parity checks for each digit
    unsatis_eq_count = H.'*eq_sol;
    % finding the maximum number of unsatisfied parity checks
    max_unsatis = max(unsatis_eq_count);
    % flipping the digits which have the maximum number of unsatisfied
    %parity checks associated with them

    rcvd_wd = rcvd_wd + (unsatis_eq_count == max_unsatis);
    rcvd_wd = mod(rcvd_wd,2);

    % reevaluating the parity checks
    eq_sol = mod((H*rcvd_wd),2);

    end %while
if( (sum(eq_sol)>0) )
    success = 0;
    if(~pass_flipped)
       cdwd = rcvd_wd_orig;
    else
       cdwd = rcvd_wd;
    end
else
    success = 1;
```

```
    cdwd = rcvd_wd;
end
```

ldpc_h2g.c

Purpose	Computes the generator matrix of the LDPC code in systematic form given a non-systematic parity check matrix in non-systematic LDPC form.
Syntax	[H_OUT, G_OUT] = ldpc_h2g(H_IN);

Description

H_OUT: Reordered parity-check matrix of the LDPC code in regular or irregular LDPC form
G_OUT: Generator matrix for the LDPC code in systematic form. This generator matrix corresponds to the reordered parity-check matrix H_OUT
H_IN: Parity-check matrix of LDPC code in non-systematic LDPC form (regular or irregular) whose generator matrix needs to be computed

The input parity-check matrix is not in systematic form. This matrix is converted to systematic form via Gaussian elimination on rows and column reordering. Once the parity-check matrix is in systematic form, the generator matrix can be easily computed (refer Wicker [75]). The gaussian elimination operations on the rows of the systematic parity-check matrix are reversed to give a parity-check matrix in regular or irregular LDPC form that now corresponds to the computed systematic generator matrix. Program written by Igor Kozintsev [40].

Program code

```
/* Author : Igor Kozintsev   igor@ifp.uiuc.edu
   Please let me know if you find bugs in this code (I did test
   it but I still have some doubts). All other comments are welcome
   too :) !
   I use a simple algorithm to invert H.
   We convert H to [I | A]
                   [junk ]
   using column reodering and row operations (junk - a few rows of H
   which are linearly dependent on the previous ones)
   G is then found as G = [A'|I]
   G is stored as array of doubles in Matlab which is very inefficient.
   Internal representation in this programm is unsigned char. Please modify
```

```
   the part which writes G if you wish.
   */
#include <math.h> #include "mex.h"

/* Input Arguments: tentative H matrix*/ #define H_IN     prhs[0]
/* Output Arguments: final matrices*/ #define H_OUT   plhs[0]
#define G_OUT   plhs[1]

void mexFunction(
                int nlhs,        mxArray *plhs[],
                int nrhs, const mxArray *prhs[]
        )
{
  unsigned char **HH, **GG;
  int ii, jj, *ir, *jc, rdep, tmp, d;
  double *sr1, *sr2, *g;
  int N,M,K,i,j,k,kk,nz,*irs1,*jcs1, *irs2, *jcs2;

  /* Check for proper number of arguments */
  if (nrhs != 1) {
    mexErrMsgTxt("h2g requires one input arguments.");
  } else if (nlhs != 2) {
    mexErrMsgTxt("h2g requires two output arguments.");
  } else if (!mxIsSparse(H_IN)) {
    mexErrMsgTxt("h2g requires sparse H matrix.");
  }
/* read sparse matrix H */
    sr1  = mxGetPr(H_IN);
    irs1 = mxGetIr(H_IN);  /* row */
    jcs1 = mxGetJc(H_IN);  /* column */
    nz = mxGetNzmax(H_IN); /* number of nonzero elements (they are ones)*/
    M = mxGetM(H_IN);
    N = mxGetN(H_IN);

/* create working array HH[row][column]*/
    HH = (unsigned char **)mxMalloc(M*sizeof(unsigned char *));
    for(i=0 ; i<M ; i++){
      HH[i] = (unsigned char *)mxMalloc(N*sizeof(unsigned char));
    }
    for(i=0 ; i<M ; i++)
      for(j=0 ; j<N ; j++)
        HH[i][j] = 0; /* initialize all to zero */

    k=0;
    for(j=0 ; j<N ; j++) {
      for(i=0 ; i<(jcs1[j+1]-jcs1[j]) ; i++) {
        ii = irs1[k]; /* index in column j*/
        HH[ii][j] = sr1[k]; /* put  nonzeros */
        k++;
      }
```

```
    }
/* invert HH matrix here */
    /* row and column indices */
    ir = (int *)mxMalloc(M*sizeof(int));
        jc = (int *)mxMalloc(N*sizeof(int));
    for( i=0 ; i<M ; i++)
        ir[i] = i;
    for( j=0 ; j<N ; j++)
        jc[j] = j;
    /* perform Gaussian elimination on H, store reodering operations */
    rdep = 0; /* number of dependent rows in H*/
    d = 0;    /* current diagonal element */

    while( (d+rdep) < M) { /* cycle through independent rows of H */

        j = d; /* current column index along row ir[d] */
        while( (HH[ir[d]][jc[j]] == 0) && (j<(N-1)) )
            j++;            /* find first nonzero element in row i */
        if( HH[ir[d]][jc[j]] ) {/*found nonzero element. It is "1" in GF2*/


            /* swap columns */
            tmp = jc[d]; jc[d] = jc[j]; jc[j] = tmp;

            /* eliminate current column using row operations */
            for(ii=0 ; ii<M ; ii++)
                if(HH[ir[ii]][jc[d]] && (ir[ii] != ir[d]))
                    for(jj=d ; jj<N ; jj++)
                        HH[ir[ii]][jc[jj]] = (HH[ir[ii]][jc[jj]]+HH[ir[d]][jc[jj]])
                        %2;
        }
        else { /* all zeros - need to delete this row and update indices */
            rdep++; /* increase number of dependent rows */
            tmp = ir[d];
            ir[d] = ir[M-rdep];
            ir[M-rdep] = tmp;
            d--; /* no diagonal element is found */
        }
        d++; /* increase the number of diagonal elements */
    }/*while i+rdep*/
/* done inverting HH */

    K = N-M+rdep; /* true K */

/* create G matrix  G = [A'| I] if H = [I|A]*/
    GG = (unsigned char **)mxMalloc(K*sizeof(unsigned char *));
    for(i=0 ; i<K ; i++){
        GG[i] = (unsigned char *)mxMalloc(N*sizeof(unsigned char));
    }
    for(i=0 ; i<K ; i++)
```

```
            for(j=0 ; j<(N-K) ; j++) {
                tmp = (N-K+i);
                GG[i][j] = HH[ir[j]][jc[tmp]];
            }


        for(i=0 ; i<K ; i++)
            for(j=(N-K); j<N ; j++)
                if(i == (j-N+K) ) /* diagonal */
                    GG[i][j] = 1;
                else
                    GG[i][j] = 0;
/* NOTE, it is very inefficient way to store G. Change to taste!*/
    G_OUT = mxCreateDoubleMatrix(K, N, mxREAL);
    /* Assign pointers to the output matrix */
    g = mxGetPr(G_OUT);
    for(i=0 ; i<K ; i++)
        for(j=0 ; j<N; j++)
            g[i+j*K] = GG[i][j];

    H_OUT = mxCreateSparse(M,N,nz,mxREAL);
    sr2  = mxGetPr(H_OUT);
    irs2 = mxGetIr(H_OUT);   /* row */
    jcs2 = mxGetJc(H_OUT);   /* column */
    /* Write H_OUT swapping columns according to jc */
    k = 0;
    for (j=0; (j<N ); j++) {
        jcs2[j] = k;
        tmp = jcs1[jc[j]+1]-jcs1[jc[j]];
        for (i=0; i<tmp ; i++) {
            kk = jcs1[jc[j]]+i;
            sr2[k] = sr1[kk];
            irs2[k] = irs1[kk];
            k++;
        }
    }
    jcs2[N] = k;

/* free the memory */
    for( j=0 ; j<M ; j++) {
        mxFree(HH[j]);
    }
    mxFree(HH);
    mxFree(ir);
    mxFree(jc);
    for(i=0;i<K;i++){
        mxFree(GG[i]);
    }
    mxFree(GG);
    return;
}
```

SELECTED BIBLIOGRAPHY

[1] G. Battail, "On Linear Random-Like Codes", *Proceedings of the 2nd International Symposium on Turbo Codes*, Brest, France, 4-7 Sept. 2000, pp. 47-50.

[2] I. Beichl and F. Sullivan, "The Importance of Importance Sampling", *Computing in Science and Engineering*, vol. 1, No. 2, Mar./Apr. 1999, pp. 71-73.

[3] S. Benedetto and E. Biglieri, *Principles of Digital Tranmission with Wireless Applications*, Kluwer Academic, 1999.

[4] B. A. Berg, "A Brief Review of Multicanonical Simulations", in, *Proceedings of the International Conference on Multiscale Phenomena and their Simulations*, Bielefeld, 30 Sept. - 4 Oct. 1996, edited by F. Karsch, B. Monien, and H. Satz, World Scientific, Singapore, June 1997, pp. 137-146.

[5] B. A. Berg, "Algorithmic Aspects of Multicanonical Simulations", *Nuclear Physics B. Proceedings Supplement*, vol. 63, No. 3, Apr. 1998, pp. 982-984. Also available at *http://www.arxiv.org/abs/hep-lat/9708003*

[6] B. A. Berg, "Introduction to Multicanonical Monte Carlo Simulations", based on lecture given at the *Fields Institute Workshop on Monte Carlo Methods*, Oct. 1998. Available via *xxx.lanl.gov/format/condmat/9909236*, Feb. 2000.

[7] B. A. Berg and T. Neuhaus, "The multicanonical ensemble: a new approach to simulate first-order phase transitions," *Physical Review Letters*, vol. 68, No. 1, Jan. 1992, pp. 912.

[8] C. Berrou, A. Glavieux, and P. Thitimajshima, "Near Shannon Limit Error Correcting Coding and Decoding: Turbo Codes", *Proceedings of the IEEE International Conference on Communications*, Geneva, Switzerland, May 1993, pp. 1064-1070.

[9] I. F. Blake and R. C. Mullin, *The Mathematical Theory of Coding*, Academic Press, Chapter 2: Combinatorial Constructions and Coding, 1975.

[10] M. Blaum, P. G. Farrell, and H. C. A. van Tilborg, "Array Codes", Chapter 22 in *Handbook of Coding Theory*, V. S. Pless and W. C. Huffman (editors), Elsevier Science B. V., New York, NY, 1998.

[11] R. Brualdi, *Introductory Combinatorics*, 3rd Edition, Prentice Hall, Upper Saddle River, NJ, 1999.

[12] G. Caire, G. Taricco, and G. Battail, "Weight Distribution and Performance of the Iterated Product of Single-Parity-Check Codes", *Annals of Telecommunications*, vol. 50, No. 9-10, Sept.-Oct. 1995, pp. 752-761.

[13] A. R. Calderbank, E. M. Rains, P. M. Shor, N. J. A. Sloane, "Quantum error correction via codes over GF(4)", *IEEE Transactions on Information Theory*, vol. 44, No. 4, July 1998, pp. 1369–1387.

[14] T. Calinski and S. Kageyama, *Block Designs: A Randomization Approach, vol. II: Design*, Springer Verlag, Chapter 1, New York, 2003.

[15] G. Clark and J. Cain, *Error-Correction Coding for Digital Communications*, Plenum, New York, 1981.

[16] C. Cohen-Tannoudji, B. Diu, and F. Laloë, *Quantum Mechanics*, vol. I, Wiley-Interscience, 1996.

[17] T. M. Cover and J. A. Thomas, *Elements of Information Theory*, John Wiley, New York, 1991.

[18] I. B. Djordjevic and B. Vasic, "Projective Geometry Low-Density Parity-Check Codes for Ultra-Long Haul WDM High-Speed Transmission", *IEEE Photonic Technology Letters*, vol. 15, No. 5, May 2003, pp. 784-786.

[19] I. B. Djordjevic and B. Vasic, "High Code Rate Low-Density Parity-Check Codes for Optical Communication Systems", *IEEE Photonics Technology Letters*, vol. 16, No. 6, June 2004, pp. 1600-1602.

[20] J. L. Fan, "Array Codes as Low Density Parity Check Codes", *Proceedings of the 2nd International Symposium on Turbo Codes and Related Topics*, Brest, France, 4-7 Sept. 2000, pp. 543-546.

[21] P. G. Farrell, "Array Codes", in *Algebraic Coding Theory and Applications*, Editor: Giuseppe Longo, Springer Verlag, 1979, pp. 231-242.

[22] M. Ferrari and S. Bellini, "Importance Sampling Simulation of Concatenated Block Codes", *IEE Proceedings on Communications*, vol. 147, No. 5, Oct. 2000, pp. 245-251.

[23] M. P. C. Fossorier, S. Lin, and D. D. Rhee, "Maximum-likelihood decoding of linear block codes and related soft-decision decoding methods," *IEEE Transactions on Information Theory*, vol. 44, No. 7, Nov. 1998, pp. 3083-3090.

[24] B. J. Frey, *Graphical Models for Machine Learning and Digital Communication*, MIT Press, Cambridge, MA, 1998.

[25] A. Friedman, *Foundations of Modern Analysis*, Dover Publications Inc., New York, 1982.

[26] H. Futaki and T. Ohtsuki, "Performance of low-density parity-check (LDPC) coded OFDM systems," *Proceedings of the IEEE International Conference on Communications 2002*, New York, NY, 28 Apr. - 2. May, 2002, vol. 3, pp. 1696-1700.

[27] R. G. Gallager, *Low Density Parity Check Codes*, MIT Press, Cambridge, MA, 1963.

[28] N. Gisin, J. Brendel, J-D. Gautier, B. Gisin, B. Huttner, G. Ribordy, W. Tittel, and H. Zbinden, "Quantum cryptography and long distance Bell experiments: How to control decoherence", *http://www.arxiv.org/PS_cache/quant-ph/pdf/9901/9901043.pdf*

[29] D. Gottesman, "An Introduction to Quantum Error Correction", *http://arxiv.org/abs/quant-ph/0004072*, 2000.

[30] M. Grassl, "Table of quantum error-correcting codes", 2003, at *http://iaks-www.ira.uka.de/home/grassl/QECC/Cyclic/index.html*

[31] J. Gubernatis and N. Hatano, "The Multicanonical Monte Carlo Method", *Computing in Science and Engineering*, vol. 2, No. 2, Mar./Apr. 2000, pp. 95-102.

[32] J. Hagenauer, E. Offer, and L. Papke, "Iterative decoding of binary block and convolutional codes", *IEEE Transactions on Information Theory*, vol. 42, No. 2, Mar. 1996, pp. 429-445.

[33] R. Holzlöhner, A. Mahadevan, C. R. Menyuk, J. M. Morris, and J. Zweck, "Evaluation of the Very Low BER of FEC Codes Using Dual Adaptive Importance Sampling", *IEEE Communications Letters*, vol. 9, No. 2, Feb. 2005, pp. 163-165.

[34] R. Holzlöhner and C. R. Menyuk, "Use of multicanonical Monte Carlo simulations to obtain accurate bit error rates in optical communications systems," *Optical Letters*, vol. 28, No. 20, Oct. 2003, pp. 1894–1896.

[35] A. W. Hunt, "Hyper-Codes: High-Performance Low-Complexity Error-Correcting Codes", M. S. Thesis, Electrical Engineering, Carleton University, Ottawa, Canada, May 1998.

[36] M. C. Jeruchim, P. Balaban, and K. S. Shanmugan, *Simulation of Communication Systems: Modeling, Methodology, and Techniques*, 2nd Ed., Kluwer Academic/Plenum, New York, 2000.

[37] S. J. Johnson and S. R. Weller, "Codes for Iterative Decoding from Partial Geometries", *Proceedings of IEEE International Symposium on Information Theory*, Lausanne, Switzerland, 30 June - 5 July 2002, pp. 310.

[38] Y. Kou, S. Lin, and M. P. C. Fossorier, "Low-Density Parity-Check Codes Based on Finite Geometries: A Rediscovery and New Results", *IEEE Transaction on Information Theory*, vol. 47, No. 7, Nov. 2001, pp. 2711-2736.

[39] Y. Kou, S. Lin, M. Fossorier, "Construction of Low Density Parity Check Codes: A Geometric Approach", *Proceedings of the 2nd International Symposium on Turbo Codes*, Brest, France, 4-7 Sept. 2000, pp. 137-140.

[40] I. Kozintsev, "Matlab Routines for LDPC Codes", *http://www.kozintsev.net/soft.html*

[41] F. R. Kschischang, B. J. Frey and H-A. Loeliger, "Factor Graphs and the Sum-Product Algorithm", *IEEE Transactions on Information Theory*, vol. 47, No. 2, Feb. 2001, pp. 498-519.

[42] B. Lapeyre, E. Pardoux, and R. Sentis, *Introduction to Monte-Carlo Methods for Transport and Diffusion Equations*, English Translation, Oxford University Press, New York, 2003.

[43] A. O. Lima, I. T. Lima Jr., and C. R. Menyuk, "Error Estimation in Multicanonical Monte Carlo Simulations with Applications to Polarization Mode Dispersion Emulators", submitted to the *IEEE Journal of Lightwave Technology*, Mar. 2005.

[44] S. Lin and D. J. Costello, *Error Correction Coding*, 2nd Ed., Prentice Hall, Engelwood Cliffs, NJ, 2004.

[45] D. G. Luenberger, *Optimization by Vector Space Methods*, John Wiley & Sons, Inc., New York, 1969.

[46] T. Mabuchi, K. Ouch, T. Kobayashi, Y. Miyata, K. Kino, H. Agama, K. Kubo, H. Yoshida, M. Akita, and K. Motoshima, "Experimental Demonstration of Net Coding Gain of 10.1 dB using 12.4 Gb/s Block Turbo Code with 3-bit Soft Decision", *Proceedings of the Optical Fiber Conference OFC 2003*, Post-deadline papers, 23-28 Mar. 2003, pp. PD21-1 to PD21-3.

[47] D. J. C. Mackay and R. M. Neal, "Near Shannon Limit Performance of Low Density Parity Check Codes", *Electronics Letters*, vol. 33, No. 6, Mar. 1997, pp. 457-458.

[48] D. J. C. MacKay, "Encyclopedia of Sparse Graph Codes", *http://www.inference.phy.cam.ac.uk/mackay/codes/EN/C/96.3.963*

[49] D. J. C. Mackay, G. Mitchison, and P. L. McFadden, "Sparse-Graph Codes for Quantum Error-Correction", *IEEE Transactions on Information Theory*, vo. 50, No. 10, Oct. 2004, pp. 2315-2330, and at *http://arxiv.org/abs/quant-ph/0304161*, 2003.

[50] A. Mahadevan and J. M. Morris, "On RCD SPC Codes as LDPC Codes Based on Arrays and Their Equivalence to Some Codes Constructed from Euclidean Geometries and Partial BIBDs", Technical Report No.: CSPL

TR: 2002-1, June 2002. Communications and Signal Processing Laboratory, CSEE Dept, UMBC, Catonsville, MD 21250. Available at *http://userpages.umbc.edu/~amahad1/Publications/TR2002_1.pdf*

[51] A. Mahadevan and J. M. Morris, "On FEC Code Performance Improvement Comparisons between the BSC and the BSC/E under Bounded Distance Decoding", *Proceedings of the Eighth Canadian Workshop on Information Theory*, Waterloo, Ontario, 18-21 May 2003, pp. 52-55.

[52] A. Mahadevan, "On LDPC Codes for ADSL", M. S. Thesis, Electrical Engineering, University of Maryland Baltimore County, Catonsville, MD 21250, Dec. 2001.

[53] A. Mahadevan and J. M. Morris, "On Quantum Codes from Weakly-Dual FEC Codes Based on Diagonals on Arrays", *Proceedings of the 2005 Canadian Workshop on Information Theory*, Montréal, Canada, 5-8 June 2005.

[54] A. Mahadevan and J. M. Morris, "SNR-Invariant Importance Sampling for Hard-Decision Decoding Performance of Linear Block Codes", submitted to the *IEEE Transactions on Communications*, Oct. 2004.

[55] W. Martin, "The WEF for a class of regular LDPC Codes: The RCD Array Codes", PhD Dissertation, CSEE Dept., University of Maryland Baltimore County (UMBC), Catonsville, MD 21250, 2003.

[56] W. R. Martin and J. M. Morris, "The RCD Array Code is a Weakly Random-like Code", Proceedings of the 3rd International Symposium on

Turbo Codes and Related Topics, Brest, France, 1-5 Sept. 2003, pp. 351-354.

[57] N. Metropolis, A. Rosenbluth, M. Rosenbluth, A. Teller, E. Teller, "Equation of state calculations by fast computing machines," *Journal Chemical Physics*, vol. 21, No. 6, June 1953, pp. 1087–1092.

[58] A. M. Mood and F. A. Graybill, *Introduction to the Theory of Statistics*, Ch. 11, 2nd Ed., McGraw-Hill, New York, 1963.

[59] M. A. Nielsen and I. L. Chuang, *Quantum Computation and Quantum Information*, Cambridge University Press, Cambridge, 2000.

[60] A. Papoulis, *Probability, Random Variables, and Stochastic Processes*, 3rd Edition, McGraw-Hill, New York, 1991.

[61] J. Preskill, "Quantum Error Correction," Caltech Lecture Notes, Physics 219, *Quantum Computation*, Ch. 7, *http://www.theory.caltech.edu/people/preskill/ph229/#lecture*, 2002.

[62] R. Pyndiah and P. Adde, "Performance of High Code Rate BTC for non-traditional applications", *Proceeding of the 3rd International Symposium on Turbo Codes and Related Topics*, Brest, France, 1-5 Sept. 2003, pp. 157-160.

[63] T. J. Richardson, M. A. Shokrollahi, and R. L. Urbanke, "Design of Capacity-Approaching Irregular Low Density Parity Check Codes", *IEEE Transactions on Information Theory*, vol. 47, No. 2, Feb. 2001, pp. 619-637.

[64] T. J. Richardson, R. L. Urbanke, "Efficient Encoding of Low Density Parity Check Codes", *IEEE Transactions on Information Theory*, vol. 47, No. 2, Feb. 2001, pp. 638-656.

[65] T. J. Richardson and R. L. Urbanke, "The Capacity of Low Density Parity Check Codes under Message Passing Decoding", *IEEE Transactions on Information Theory*, vol. 47, No. 2, Feb. 2001, pp. 599-618.

[66] J. S. Sadowsky, "A New Method for Viterbi Decoder Simulation Using Importance Sampling", *IEEE Transactions on Communications*, vol. 38, No. 9, Sept. 1990, pp. 1341-1351.

[67] P. Scherk and R. Lingenberg, *Mathematical Expositions No. 20, Rudiments of Plane Affine Geometry*, University of Toronto Press, 1975.

[68] C. E. Shannon, *A Mathematical Theory of Communication*, 50th Anniversary Edition, Bell Labs, Lucent Technologies, May 1998.

[69] H. Steendam and M. Moeneclaey, "ML-Performance of low-density parity check codes", *Proceedings of the BENELUX-IT, 23rd Symposium on Information Theory in the Benelux*, Louvain-la-Neuve, Belgium, May 2002, pp. 75–78.

[70] B. Vasic, "High-rate low-density parity check codes based on anti-Pasch affine geometries", *Proceedings of the IEEE International Conference on Communications (ICC 2002)*, New York, NY, 28 Apr. - 2 May 2002, vol. 3, pp. 1332-1336.

[71] B. Vasic, personal communication, Feb 15 2003.

[72] J. Wang, "Performance Bounds and Design Rules for Product Codes with Hyper-Diagonal Parity", M. S. Thesis, Electrical Engineering, Washington State University, Pullman, WA, Aug. 2000, available at *http://www.eecs.wsu.edu/~belzer/TPC_thesis.pdf*

[73] N. Wiberg, H.-A. Loeliger, R. Kötter, "Codes and Iterative Decoding on General Graphs", *European Transactions on Telecommunications*, vol. 6, Sept./Oct. 1995, pp. 513-525.

[74] N. Wiberg, "Codes and Decoding on General Graphs", PhD. Dissertation, Department of Electrical Engineering, Linköping University, Sweden, 1996.

[75] S. Wicker, *Error Control Systems for Digital Communication and Storage*, Prentice Hall, Upper Saddle River, NJ, 1995.

[76] X. Wu and S. G. Wilson, "Importance Sampling for Estimating Decoder Performance", *Proceedings of the 29th Conference on Information Sciences and Systems*, Baltimore, MD, 22-24 Mar. 1995, pp. 402-407.

[77] B. Xia and W. E. Ryan, "On importance sampling for linear block codes", *Proceedings of the International Conference on Communications 2003*, Anchorage, AK, 11-15 May 2003, pp. 2904-2908.

[78] J. Xu, H. Tang, Y. Kou, S. Lin, and K. Abdel-Ghaffar, "A Geometric Approach to the Construction of Gallager Codes", *Proceedings of the Conference on Information Sciences and Systems (CISS 2002)*, Princeton, NJ, 20-22 Mar. 2002, pp. 552-555.

[79] T. Yamane and Y. Katayama, "Bit Error Rate Analysis on Iterative Two-Stage Decoding of Two-Dimensional Codes by Importance Sampling", *Proceedings of the International Conference on Communications 2003*, Anchorage, AK, 11-15 May 2003, pp. 3140-3144.

www.ingramcontent.com/pod-product-compliance
Lightning Source LLC
Chambersburg PA
CBHW071359050326
40689CB00010B/1697